Microsoft

2210B: Expert Track: Updating Systems Engineer Skills from Microsoft® Windows® 2000 to Windows Server™ 2003

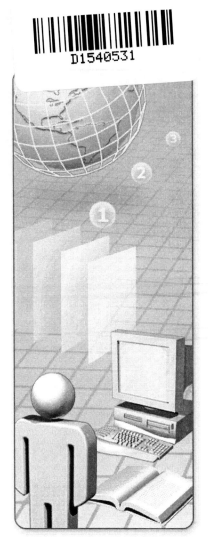

Workshop: 2210B
Part Number: X10-00813
Released: 10/2003

END-USER LICENSE AGREEMENT FOR OFFICIAL MICROSOFT LEARNING PRODUCTS – STUDENT EDITION

PLEASE READ THIS END-USER LICENSE AGREEMENT ("EULA") CAREFULLY. BY USING THE MATERIALS AND/OR USING OR INSTALLING THE SOFTWARE THAT ACCOMPANIES THIS EULA (COLLECTIVELY, THE "LICENSED CONTENT"), YOU AGREE TO THE TERMS OF THIS EULA. IF YOU DO NOT AGREE, DO NOT USE THE LICENSED CONTENT.

1.   **GENERAL.** This EULA is a legal agreement between you (either an individual or a single entity) and Microsoft Corporation ("Microsoft"). This EULA governs the Licensed Content, which includes computer software (including online and electronic documentation), training materials, and any other associated media and printed materials. This EULA applies to updates, supplements, add-on components, and Internet-based services components of the Licensed Content that Microsoft may provide or make available to you unless Microsoft provides other terms with the update, supplement, add-on component, or Internet-based services component. Microsoft reserves the right to discontinue any Internet-based services provided to you or made available to you through the use of the Licensed Content. This EULA also governs any product support services relating to the Licensed Content except as may be included in another agreement between you and Microsoft. An amendment or addendum to this EULA may accompany the Licensed Content.

2.   **GENERAL GRANT OF LICENSE.** Microsoft grants you the following rights, conditioned on your compliance with all the terms and conditions of this EULA. Microsoft grants you a limited, non-exclusive, royalty-free license to install and use the Licensed Content solely in conjunction with your participation as a student in an Authorized Training Session (as defined below). You may install and use one copy of the software on a single computer, device, workstation, terminal, or other digital electronic or analog device ("Device"). You may make a second copy of the software and install it on a portable Device for the exclusive use of the person who is the primary user of the first copy of the software. A license for the software may not be shared for use by multiple end users. An "Authorized Training Session" means a training session conducted at a Microsoft Certified Technical Education Center, an IT Academy, via a Microsoft Certified Partner, or such other entity as Microsoft may designate from time to time in writing, by a Microsoft Certified Trainer (for more information on these entities, please visit www.microsoft.com). WITHOUT LIMITING THE FOREGOING, COPYING OR REPRODUCTION OF THE LICENSED CONTENT TO ANY SERVER OR LOCATION FOR FURTHER REPRODUCTION OR REDISTRIBUTION IS EXPRESSLY PROHIBITED.

3.   **DESCRIPTION OF OTHER RIGHTS AND LICENSE LIMITATIONS**

   3.1   *Use of Documentation and Printed Training Materials.*

         3.1.1   The documents and related graphics included in the Licensed Content may include technical inaccuracies or typographical errors. Changes are periodically made to the content. Microsoft may make improvements and/or changes in any of the components of the Licensed Content at any time without notice. The names of companies, products, people, characters and/or data mentioned in the Licensed Content may be fictitious and are in no way intended to represent any real individual, company, product or event, unless otherwise noted.

         3.1.2   Microsoft grants you the right to reproduce portions of documents (such as student workbooks, white papers, press releases, datasheets and FAQs) (the "Documents") provided with the Licensed Content. You may not print any book (either electronic or print version) in its entirety. If you choose to reproduce Documents, you agree that: (a) use of such printed Documents will be solely in conjunction with your personal training use; (b) the Documents will not republished or posted on any network computer or broadcast in any media; (c) any reproduction will include either the Document's original copyright notice or a copyright notice to Microsoft's benefit substantially in the format provided below; and (d) to comply with all terms and conditions of this EULA. In addition, no modifications may made to any Document.

         Form of Notice:

         © 2003. Reprinted with permission by Microsoft Corporation. All rights reserved.

         Microsoft and Windows are either registered trademarks or trademarks of Microsoft Corporation in the US and/or other countries. Other product and company names mentioned herein may be the trademarks of their respective owners.

   3.2   *Use of Media Elements.* The Licensed Content may include certain photographs, clip art, animations, sounds, music, and video clips (together "Media Elements"). You may not modify these Media Elements.

   3.3   *Use of Sample Code.* In the event that the Licensed Content includes sample code in source or object format ("Sample Code"), Microsoft grants you a limited, non-exclusive, royalty-free license to use, copy and modify the Sample Code; if you elect to exercise the foregoing rights, you agree to comply with all other terms and conditions of this EULA, including without limitation Sections 3.4, 3.5, and 6.

   3.4   *Permitted Modifications.* In the event that you exercise any rights provided under this EULA to create modifications of the Licensed Content, you agree that any such modifications: (a) will not be used for providing training where a fee is charged in public or private classes; (b) indemnify, hold harmless, and defend Microsoft from and against any claims or lawsuits, including attorneys' fees, which arise from or result from your use of any modified version of the Licensed Content; and (c) not to transfer or assign any rights to any modified version of the Licensed Content to any third party without the express written permission of Microsoft.

3.5   *Reproduction/Redistribution Licensed Content.* Except as expressly provided in this EULA, you may not reproduce or distribute the Licensed Content or any portion thereof (including any permitted modifications) to any third parties without the express written permission of Microsoft.

4.   **RESERVATION OF RIGHTS AND OWNERSHIP.** Microsoft reserves all rights not expressly granted to you in this EULA. The Licensed Content is protected by copyright and other intellectual property laws and treaties. Microsoft or its suppliers own the title, copyright, and other intellectual property rights in the Licensed Content. You may not remove or obscure any copyright, trademark or patent notices that appear on the Licensed Content, or any components thereof, as delivered to you. **The Licensed Content is licensed, not sold.**

5.   **LIMITATIONS ON REVERSE ENGINEERING, DECOMPILATION, AND DISASSEMBLY.** You may not reverse engineer, decompile, or disassemble the Software or Media Elements, except and only to the extent that such activity is expressly permitted by applicable law notwithstanding this limitation.

6.   **LIMITATIONS ON SALE, RENTAL, ETC. AND CERTAIN ASSIGNMENTS.** You may not provide commercial hosting services with, sell, rent, lease, lend, sublicense, or assign copies of the Licensed Content, or any portion thereof (including any permitted modifications thereof) on a stand-alone basis or as part of any collection, product or service.

7.   **CONSENT TO USE OF DATA.** You agree that Microsoft and its affiliates may collect and use technical information gathered as part of the product support services provided to you, if any, related to the Licensed Content. Microsoft may use this information solely to improve our products or to provide customized services or technologies to you and will not disclose this information in a form that personally identifies you.

8.   **LINKS TO THIRD PARTY SITES.** You may link to third party sites through the use of the Licensed Content. The third party sites are not under the control of Microsoft, and Microsoft is not responsible for the contents of any third party sites, any links contained in third party sites, or any changes or updates to third party sites. Microsoft is not responsible for webcasting or any other form of transmission received from any third party sites. Microsoft is providing these links to third party sites to you only as a convenience, and the inclusion of any link does not imply an endorsement by Microsoft of the third party site.

9.   **ADDITIONAL LICENSED CONTENT/SERVICES.** This EULA applies to updates, supplements, add-on components, or Internet-based services components, of the Licensed Content that Microsoft may provide to you or make available to you after the date you obtain your initial copy of the Licensed Content, unless we provide other terms along with the update, supplement, add-on component, or Internet-based services component. Microsoft reserves the right to discontinue any Internet-based services provided to you or made available to you through the use of the Licensed Content.

10.   **U.S. GOVERNMENT LICENSE RIGHTS**. All software provided to the U.S. Government pursuant to solicitations issued on or after December 1, 1995 is provided with the commercial license rights and restrictions described elsewhere herein. All software provided to the U.S. Government pursuant to solicitations issued prior to December 1, 1995 is provided with "Restricted Rights" as provided for in FAR, 48 CFR 52.227-14 (JUNE 1987) or DFAR, 48 CFR 252.227-7013 (OCT 1988), as applicable.

11.   **EXPORT RESTRICTIONS**. You acknowledge that the Licensed Content is subject to U.S. export jurisdiction. You agree to comply with all applicable international and national laws that apply to the Licensed Content, including the U.S. Export Administration Regulations, as well as end-user, end-use, and destination restrictions issued by U.S. and other governments. For additional information see <http://www.microsoft.com/exporting/>.

12.   **TRANSFER.** The initial user of the Licensed Content may make a one-time permanent transfer of this EULA and Licensed Content to another end user, provided the initial user retains no copies of the Licensed Content. The transfer may not be an indirect transfer, such as a consignment. Prior to the transfer, the end user receiving the Licensed Content must agree to all the EULA terms.

13.   **"NOT FOR RESALE" LICENSED CONTENT.** Licensed Content identified as "Not For Resale" or "NFR," may not be sold or otherwise transferred for value, or used for any purpose other than demonstration, test or evaluation.

14.   **TERMINATION.** Without prejudice to any other rights, Microsoft may terminate this EULA if you fail to comply with the terms and conditions of this EULA. In such event, you must destroy all copies of the Licensed Content and all of its component parts.

15.   <u>DISCLAIMER OF WARRANTIES.</u> **TO THE MAXIMUM EXTENT PERMITTED BY APPLICABLE LAW, MICROSOFT AND ITS SUPPLIERS PROVIDE THE LICENSED CONTENT AND SUPPORT SERVICES (IF ANY)** *AS IS AND WITH ALL FAULTS,* **AND MICROSOFT AND ITS SUPPLIERS HEREBY DISCLAIM ALL OTHER WARRANTIES AND CONDITIONS, WHETHER EXPRESS, IMPLIED OR STATUTORY, INCLUDING, BUT NOT LIMITED TO, ANY (IF ANY) IMPLIED WARRANTIES, DUTIES OR CONDITIONS OF MERCHANTABILITY, OF FITNESS FOR A PARTICULAR PURPOSE, OF RELIABILITY OR AVAILABILITY, OF ACCURACY OR COMPLETENESS OF RESPONSES, OF RESULTS, OF WORKMANLIKE EFFORT, OF LACK OF VIRUSES, AND OF LACK OF NEGLIGENCE, ALL WITH REGARD TO THE LICENSED CONTENT, AND THE PROVISION OF OR FAILURE TO PROVIDE SUPPORT OR OTHER SERVICES, INFORMATION, SOFTWARE, AND RELATED CONTENT THROUGH THE LICENSED CONTENT, OR OTHERWISE ARISING OUT OF THE USE OF THE LICENSED CONTENT. ALSO, THERE IS NO WARRANTY OR CONDITION OF TITLE, QUIET ENJOYMENT, QUIET POSSESSION, CORRESPONDENCE TO DESCRIPTION OR NON-INFRINGEMENT WITH REGARD TO THE LICENSED CONTENT. THE ENTIRE RISK AS TO THE QUALITY, OR ARISING OUT OF THE USE OR PERFORMANCE OF THE LICENSED CONTENT, AND ANY SUPPORT SERVICES, REMAINS WITH YOU.**

16.   <u>EXCLUSION OF INCIDENTAL, CONSEQUENTIAL AND CERTAIN OTHER DAMAGES.</u> **TO THE MAXIMUM EXTENT PERMITTED BY APPLICABLE LAW, IN NO EVENT SHALL MICROSOFT OR ITS SUPPLIERS BE LIABLE FOR ANY SPECIAL, INCIDENTAL, PUNITIVE, INDIRECT, OR CONSEQUENTIAL DAMAGES WHATSOEVER (INCLUDING, BUT NOT**

LIMITED TO, DAMAGES FOR LOSS OF PROFITS OR CONFIDENTIAL OR OTHER INFORMATION, FOR BUSINESS INTERRUPTION, FOR PERSONAL INJURY, FOR LOSS OF PRIVACY, FOR FAILURE TO MEET ANY DUTY INCLUDING OF GOOD FAITH OR OF REASONABLE CARE, FOR NEGLIGENCE, AND FOR ANY OTHER PECUNIARY OR OTHER LOSS WHATSOEVER) ARISING OUT OF OR IN ANY WAY RELATED TO THE USE OF OR INABILITY TO USE THE LICENSED CONTENT, THE PROVISION OF OR FAILURE TO PROVIDE SUPPORT OR OTHER SERVICES, INFORMATION, SOFTWARE, AND RELATED CONTENT THROUGH THE LICENSED CONTENT, OR OTHERWISE ARISING OUT OF THE USE OF THE LICENSED CONTENT, OR OTHERWISE UNDER OR IN CONNECTION WITH ANY PROVISION OF THIS EULA, EVEN IN THE EVENT OF THE FAULT, TORT (INCLUDING NEGLIGENCE), MISREPRESENTATION, STRICT LIABILITY, BREACH OF CONTRACT OR BREACH OF WARRANTY OF MICROSOFT OR ANY SUPPLIER, AND EVEN IF MICROSOFT OR ANY SUPPLIER HAS BEEN ADVISED OF THE POSSIBILITY OF SUCH DAMAGES. BECAUSE SOME STATES/JURISDICTIONS DO NOT ALLOW THE EXCLUSION OR LIMITATION OF LIABILITY FOR CONSEQUENTIAL OR INCIDENTAL DAMAGES, THE ABOVE LIMITATION MAY NOT APPLY TO YOU.

17. <u>LIMITATION OF LIABILITY AND REMEDIES.</u> NOTWITHSTANDING ANY DAMAGES THAT YOU MIGHT INCUR FOR ANY REASON WHATSOEVER (INCLUDING, WITHOUT LIMITATION, ALL DAMAGES REFERENCED HEREIN AND ALL DIRECT OR GENERAL DAMAGES IN CONTRACT OR ANYTHING ELSE), THE ENTIRE LIABILITY OF MICROSOFT AND ANY OF ITS SUPPLIERS UNDER ANY PROVISION OF THIS EULA AND YOUR EXCLUSIVE REMEDY HEREUNDER SHALL BE LIMITED TO THE GREATER OF THE ACTUAL DAMAGES YOU INCUR IN REASONABLE RELIANCE ON THE LICENSED CONTENT UP TO THE AMOUNT ACTUALLY PAID BY YOU FOR THE LICENSED CONTENT OR US$5.00. THE FOREGOING LIMITATIONS, EXCLUSIONS AND DISCLAIMERS SHALL APPLY TO THE MAXIMUM EXTENT PERMITTED BY APPLICABLE LAW, EVEN IF ANY REMEDY FAILS ITS ESSENTIAL PURPOSE.

18. **APPLICABLE LAW.** If you acquired this Licensed Content in the United States, this EULA is governed by the laws of the State of Washington. If you acquired this Licensed Content in Canada, unless expressly prohibited by local law, this EULA is governed by the laws in force in the Province of Ontario, Canada; and, in respect of any dispute which may arise hereunder, you consent to the jurisdiction of the federal and provincial courts sitting in Toronto, Ontario. If you acquired this Licensed Content in the European Union, Iceland, Norway, or Switzerland, then local law applies. If you acquired this Licensed Content in any other country, then local law may apply.

19. **ENTIRE AGREEMENT; SEVERABILITY.** This EULA (including any addendum or amendment to this EULA which is included with the Licensed Content) are the entire agreement between you and Microsoft relating to the Licensed Content and the support services (if any) and they supersede all prior or contemporaneous oral or written communications, proposals and representations with respect to the Licensed Content or any other subject matter covered by this EULA. To the extent the terms of any Microsoft policies or programs for support services conflict with the terms of this EULA, the terms of this EULA shall control. If any provision of this EULA is held to be void, invalid, unenforceable or illegal, the other provisions shall continue in full force and effect.

Should you have any questions concerning this EULA, or if you desire to contact Microsoft for any reason, please use the address information enclosed in this Licensed Content to contact the Microsoft subsidiary serving your country or visit Microsoft on the World Wide Web at http://www.microsoft.com.

Si vous avez acquis votre Contenu Sous Licence Microsoft au CANADA :

DÉNI DE GARANTIES. Dans la mesure maximale permise par les lois applicables, le Contenu Sous Licence et les services de soutien technique (le cas échéant) sont fournis *TELS QUELS ET AVEC TOUS LES DÉFAUTS* par Microsoft et ses fournisseurs, lesquels par les présentes dénient toutes autres garanties et conditions expresses, implicites ou en vertu de la loi, notamment, mais sans limitation, (le cas échéant) les garanties, devoirs ou conditions implicites de qualité marchande, d'adaptation à une fin usage particulière, de fiabilité ou de disponibilité, d'exactitude ou d'exhaustivité des réponses, des résultats, des efforts déployés selon les règles de l'art, d'absence de virus et d'absence de négligence, le tout à l'égard du Contenu Sous Licence et de la prestation des services de soutien technique ou de l'omission de la 'une telle prestation des services de soutien technique ou à l'égard de la fourniture ou de l'omission de la fourniture de tous autres services, renseignements, Contenus Sous Licence, et contenu qui s'y rapporte grâce au Contenu Sous Licence ou provenant autrement de l'utilisation du Contenu Sous Licence. PAR AILLEURS, IL N'Y A AUCUNE GARANTIE OU CONDITION QUANT AU TITRE DE PROPRIÉTÉ, À LA JOUISSANCE OU LA POSSESSION PAISIBLE, À LA CONCORDANCE À UNE DESCRIPTION NI QUANT À UNE ABSENCE DE CONTREFAÇON CONCERNANT LE CONTENU SOUS LICENCE.

<u>EXCLUSION DES DOMMAGES ACCESSOIRES, INDIRECTS ET DE CERTAINS AUTRES DOMMAGES.</u> DANS LA MESURE MAXIMALE PERMISE PAR LES LOIS APPLICABLES, EN AUCUN CAS MICROSOFT OU SES FOURNISSEURS NE SERONT RESPONSABLES DES DOMMAGES SPÉCIAUX, CONSÉCUTIFS, ACCESSOIRES OU INDIRECTS DE QUELQUE NATURE QUE CE SOIT (NOTAMMENT, LES DOMMAGES À L'ÉGARD DU MANQUE À GAGNER OU DE LA DIVULGATION DE RENSEIGNEMENTS CONFIDENTIELS OU AUTRES, DE LA PERTE D'EXPLOITATION, DE BLESSURES CORPORELLES, DE LA VIOLATION DE LA VIE PRIVÉE, DE L'OMISSION DE REMPLIR TOUT DEVOIR, Y COMPRIS D'AGIR DE BONNE FOI OU D'EXERCER UN SOIN RAISONNABLE, DE LA NÉGLIGENCE ET DE TOUTE AUTRE PERTE PÉCUNIAIRE OU AUTRE PERTE

DE QUELQUE NATURE QUE CE SOIT) SE RAPPORTANT DE QUELQUE MANIÈRE QUE CE SOIT À L'UTILISATION DU CONTENU SOUS LICENCE OU À L'INCAPACITÉ DE S'EN SERVIR, À LA PRESTATION OU À L'OMISSION DE LA 'UNE TELLE PRESTATION DE SERVICES DE SOUTIEN TECHNIQUE OU À LA FOURNITURE OU À L'OMISSION DE LA FOURNITURE DE TOUS AUTRES SERVICES, RENSEIGNEMENTS, CONTENUS SOUS LICENCE, ET CONTENU QUI S'Y RAPPORTE GRÂCE AU CONTENU SOUS LICENCE OU PROVENANT AUTREMENT DE L'UTILISATION DU CONTENU SOUS LICENCE OU AUTREMENT AUX TERMES DE TOUTE DISPOSITION DE LA U PRÉSENTE CONVENTION EULA OU RELATIVEMENT À UNE TELLE DISPOSITION, MÊME EN CAS DE FAUTE, DE DÉLIT CIVIL (Y COMPRIS LA NÉGLIGENCE), DE RESPONSABILITÉ STRICTE, DE VIOLATION DE CONTRAT OU DE VIOLATION DE GARANTIE DE MICROSOFT OU DE TOUT FOURNISSEUR ET MÊME SI MICROSOFT OU TOUT FOURNISSEUR A ÉTÉ AVISÉ DE LA POSSIBILITÉ DE TELS DOMMAGES.

LIMITATION DE RESPONSABILITÉ ET RECOURS.   MALGRÉ LES DOMMAGES QUE VOUS PUISSIEZ SUBIR POUR QUELQUE MOTIF QUE CE SOIT (NOTAMMENT, MAIS SANS LIMITATION, TOUS LES DOMMAGES SUSMENTIONNÉS ET TOUS LES DOMMAGES DIRECTS OU GÉNÉRAUX OU AUTRES), LA SEULE RESPONSABILITÉ 'OBLIGATION INTÉGRALE DE MICROSOFT ET DE L'UN OU L'AUTRE DE SES FOURNISSEURS AUX TERMES DE TOUTE DISPOSITION DEU LA PRÉSENTE CONVENTION EULA ET VOTRE RECOURS EXCLUSIF À L'ÉGARD DE TOUT CE QUI PRÉCÈDE SE LIMITE AU PLUS ÉLEVÉ ENTRE LES MONTANTS SUIVANTS : LE MONTANT QUE VOUS AVEZ RÉELLEMENT PAYÉ POUR LE CONTENU SOUS LICENCE OU 5,00 $US. LES LIMITES, EXCLUSIONS ET DÉNIS QUI PRÉCÈDENT (Y COMPRIS LES CLAUSES CI-DESSUS), S'APPLIQUENT DANS LA MESURE MAXIMALE PERMISE PAR LES LOIS APPLICABLES, MÊME SI TOUT RECOURS N'ATTEINT PAS SON BUT ESSENTIEL.

À moins que cela ne soit prohibé par le droit local applicable, la présente Convention est régie par les lois de la province d'Ontario, Canada. Vous consentez Chacune des parties à la présente reconnaît irrévocablement à la compétence des tribunaux fédéraux et provinciaux siégeant à Toronto, dans de la province d'Ontario et consent à instituer tout litige qui pourrait découler de la présente auprès des tribunaux situés dans le district judiciaire de York, province d'Ontario.

Au cas où vous auriez des questions concernant cette licence ou que vous désiriez vous mettre en rapport avec Microsoft pour quelque raison que ce soit, veuillez utiliser l'information contenue dans le Contenu Sous Licence pour contacter la filiale de succursale Microsoft desservant votre pays, dont l'adresse est fournie dans ce produit, ou visitez écrivez à : Microsoft sur le World Wide Web à http://www.microsoft.com

Contents

About This Workshop

This section provides you with a brief description of the workshop, audience, suggested prerequisites, and workshop objectives.

Description

This 200-300 level, three-day, instructor-led workshop provides Microsoft® Windows® 2000 Microsoft Certified Systems Engineers (MCSEs) with a way to update both their skill sets and certifications by providing a quick and effective way to obtain those skills and prepare for Microsoft Windows Server™ 2003 certification exams.

This discovery-based workshop consists primarily of labs that provide hands-on experience focused exclusively on the skills and objectives that align with Exam 70-296: *Planning, Implementing, and Maintaining a Microsoft Windows Server 2003 Environment for MCSEs Certified on Windows 2000*.

Audience

This workshop is intended for Windows 2000 MCSEs or individuals with equivalent knowledge. The *track jumper*, someone in the middle of getting his or her MCSE, is not the intended audience for this workshop. The target audience profile meets the criteria that the following table outlines.

Job role	Skill level	Product and technology experience	Preferred learning style
MCSE	300	Experience supporting and managing a Windows 2000 Active Directory® directory service environment	Instructor-led classroom training - or - online/self-paced training

Student prerequisites

This workshop requires that students meet the following prerequisites:

- Hold Windows 2000 MCSE or have equivalent knowledge and skills
- Have completed Workshop 2209: *Expert Track: Updating System Administrator Skills from Microsoft Windows 2000 to Windows Server 2003*

Workshop objectives

After completing this workshop, the student will be able to:

- Plan a Domain Name System (DNS) strategy for an enterprise organization.
- Plan for an implementation of Active Directory and for Active Directory replication.
- Implement Active Directory and DNS.
- Troubleshoot Transmission Control Protocol/Internet Protocol (TCP/IP), name resolution, and Group Policy.
- Plan and implement cross-forest trust and new security options.
- Use Active Directory and Group Policy to deploy and restrict software.
- Use Active Directory and Group Policy to set advanced security settings.
- Plan and implement secure routing and remote access.

Student Materials Compact Disc Contents

The Student Materials compact disc contains the following files and folders:

- *Autorun.exe*. When the compact disc is inserted into the CD-ROM drive, or when you double-click the **Autorun.exe** file, this file opens the compact disc and allows you to browse the Student Materials compact disc.

- *Autorun.inf*. When the compact disc is inserted into the compact disc drive, this file opens Autorun.exe.

- *Default.htm*. This file opens the Student Materials Web page. It provides you with resources pertaining to this workshop, including additional reading, review and lab answers, lab files, multimedia presentations, and workshop-related Web sites.

- *Readme.txt*. This file explains how to install the software for viewing the Student Materials compact disc and its contents and how to open the Student Materials Web page.

- *Addread*. This folder contains additional reading pertaining to this workshop.

- *Examread*. This folder contains exam readiness reviews, which explain the objectives from the exam and provide references to study aids

- *Flash*. This folder contains the installer for the Macromedia Flash 6.0 browser plug-in.

- *Fonts*. This folder contains fonts that may be required to view the Microsoft Word documents that are included with this workshop.

- *Media*. This folder contains files that are used in multimedia presentations for this workshop.

- *Mplayer*. This folder contains the setup file to install Microsoft Windows Media™ Player.

- *Toolkit*. This folder contains the files for the Resource Toolkit.

- *Webfiles*. This folder contains the files that are required to view the workshop Web page. To open the Web page, open Windows Explorer, and in the root directory of the compact disc, double-click **Default.htm** or **Autorun.exe**.

- *Wordview*. This folder contains the Word Viewer that is used to view any Word document (.doc) files that are included on the compact disc.

Document Conventions

The following conventions are used in workshop materials to distinguish elements of the text.

Convention	Use
	Represents Toolbox resources available by launching the Resource Toolkit shortcut on the desktop.
Bold	Represents commands, command options, and syntax that must be typed exactly as shown. It also indicates commands on menus and buttons, dialog box titles and options, and icon and menu names.
Italic	In syntax statements or descriptive text, indicates argument names or placeholders for variable information. Italic is also used for introducing new terms, for book titles, and for emphasis in the text.
Title Capitals	Indicate domain names, user names, computer names, directory names, and folder and file names, except when specifically referring to case-sensitive names. Unless otherwise indicated, you can use lowercase letters when you type a directory name or file name in a dialog box or at a command prompt.
ALL CAPITALS	Indicate the names of keys, key sequences, and key combinations—for example, ALT+SPACEBAR.
monospace	Represents code samples or examples of screen text.
[ ]	In syntax statements, enclose optional items. For example, [*filename*] in command syntax indicates that you can choose to type a file name with the command. Type only the information within the brackets, not the brackets themselves.
{ }	In syntax statements, enclose required items. Type only the information within the braces, not the braces themselves.
|	In syntax statements, separates an either/or choice.
▶	Indicates a procedure with sequential steps.
...	In syntax statements, specifies that the preceding item may be repeated.
. . .	Represents an omitted portion of a code sample.

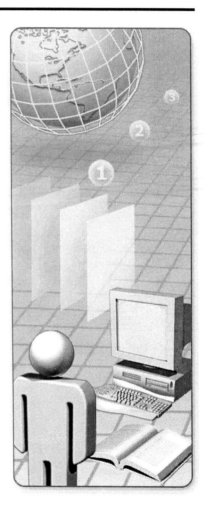

Introduction

Contents

Introduction

- Name
- Company affiliation
- Title/function
- Job responsibility
- Systems engineer experience
- Windows operating systems experience
- Expectations for the workshop

What Is a Workshop?

- **A workshop provides:**
  - Fast-paced, hands-on learning
  - A learn-as-you-go environment
- **A workshop consists of:**
  - Lecture (minimal)
  - Scenario-based labs
  - Resource Toolkit

The workshop is a fast-paced learning format that favors labs over lecture. In a workshop, lecture time is kept to a minimum to give students the opportunity to focus on hands-on, scenario-based labs. The workshop format enables students to reinforce learning by performing tasks and by problem solving.

Because lecture focuses only on the important or most difficult elements of a given topic, the labs include a Resource Toolkit that contains information like procedures, demonstrations, job aids, and other resources that are designed to give you the information you need to complete a lab. Your instructor can also answer questions to help you complete the lab. The instructor will lead discussions after the lab and review best practices.

Workshop Materials

- **Name card**
- **Student workbook**
- **Resource Toolkit**
- **Student Materials compact disc**
- **Workshop evaluation**
- **Evaluation software**

The following materials are included with your kit:

- *Name card*. Write your name on both sides of the name card.

- *Student workbook*. The student workbook contains the material covered in class, in addition to the hands-on lab exercises.

- *Resource Toolkit*. The Resource Toolkit is an online interface that contains resources that you will use in the scenario-based labs in this workshop. It includes video presentations, lab scenario information, and Toolbox resources, such as procedures and annotated screenshots, that will help you complete the labs. Most Toolbox resources are also printed in the *Resource Toolkit: Toolbox* book.

- *Student Materials compact disc*. The Student Materials compact disc contains the Web page that provides you with links to resources pertaining to this workshop, including additional readings, lab files, multimedia presentations, and workshop-related Web sites.

 Note To open the Web page, insert the Student Materials compact disc into the CD-ROM drive, and then in the root directory of the compact disc, double-click **Autorun.exe** or **Default.htm**.

- *Workshop evaluation*. To provide feedback on the workshop, training facility, and instructor, you will have the opportunity to complete an online evaluation near the end of the workshop.

  To provide additional comments or feedback on the workshop, send e-mail to support@mscourseware.com. To inquire about the Microsoft® Certified Professional program, send e-mail to mcphelp@microsoft.com.

- *Evaluation software*. An evaluation copy of the software is provided for your personal use only.

Prerequisites

- MCSE on Windows 2000 certification, or equivalent knowledge and skills
- Have completed Workshop 2209, *Expert Track: Updating Systems Administrator Skills from Microsoft Windows 2000 to Windows Server 2003*, or have equivalent knowledge and skills

---

This workshop requires that you meet the following prerequisites:

- Hold Microsoft Certified Systems Engineer (MCSE) on Microsoft® Windows® certification or possess equivalent knowledge and skills

- Have completed Workshop 2209, *Expert Track: Updating Systems Administrator Skills from Microsoft Windows 2000 to Windows Server 2003* to obtain equivalent knowledge and skills

Workshop Outline

- **Unit 1: Introduction to Performing Systems Engineer Skills in Windows Server 2003**
- **Unit 2: Planning a DNS Namespace Design**
- **Unit 3: Planning Active Directory Deployment**
- **Unit 4: Implementing DNS with Active Directory**
- **Unit 5: Troubleshooting TCP/IP, Name Resolution, and Group Policy**

Unit 1, "Introduction to Performing Systems Engineer Skills in Windows Server 2003," provides an overview of the skills that you will need to perform the systems engineer tasks in this workshop. After completing this unit, you will be able to identify the systems engineer tasks in Microsoft Windows Server™ 2003 that are new or different from the tasks performed in Windows 2000.

Unit 2, "Planning a DNS Namespace Design," discusses the new features of Domain Name System (DNS) in Windows Server 2003, such as stub zones and conditional forwarding, and focuses on planning issues .After completing this unit, you will be able to plan a DNS strategy for an enterprise organization by using the new DNS features in Windows Server 2003.

Unit 3, "Planning Active Directory Deployment," introduces new features in Microsoft Active Directory® directory service, such as universal group membership caching, link value replication (LVR), and DCAutoSite coverage. After completing this unit, you will be able to use the new features of Windows Server 2003 to plan for an implementation of Active Directory and for Active Directory replication.

Unit 4, "Implementing DNS with Active Directory," builds on the topics covered in Units 2 and 3 and focuses on implementation tasks. After completing this unit, you will be able to use the new features of Windows Server 2003 to implement DNS and Active Directory.

Unit 5, "Troubleshooting TCP/IP, Name Resolution, and Group Policy," introduces new troubleshooting tools in Windows Server 2003. After completing this unit, you will be able use these tools to troubleshoot issues regarding Transmission Control Protocol/Internet Protocol (TCP/IP), name resolution in DNS, and the application of Group Policy.

Workshop Outline *(continued)*

- Unit 6: Planning and Implementing Multiple Forests in Active Directory
- Unit 7: Using Group Policy in Windows Server 2003 to Deploy and Restrict Software
- Unit 8: Using Group Policy in Windows Server 2003 to Set Advanced Security Settings
- Unit 9: Planning and Implementing Secure Routing and Remote Access

Unit 6, "Planning and Implementing Multiple Forests in Active Directory," introduces the ability to establish cross-forest trusts in Window Server 2003. After completing this unit, you will be able to plan and implement cross-forest trust and security options in Windows Server 2003.

Unit 7, "Using Group Policy in Windows Server 2003 to Deploy and Restrict Software," discusses software restriction policies, Windows Management Instrumentation (WMI) filtering, and software distribution packages. After completing this unit, you will be able to use Active Directory and Group Policy to deploy and restrict software.

Unit 8, "Using Group Policy in Windows Server 2003 to Set Advanced Security Settings," discusses new settings in Group Policy to configure security settings for wireless networks and settings to manage the use of Encrypting File System (EFS) in an organization. After completing this unit, you will be able to use Active Directory and Group Policy to set advanced security settings.

Unit 9, "Planning and Implementing Secure Routing and Remote Access," introduces security features to protect the communication path between organizational units and access for remote users. After completing this unit, you will be able to plan and implement secure routing and remote access and troubleshoot common Internet Protocol (IP) Security (IPSec) configuration errors.

Microsoft Learning

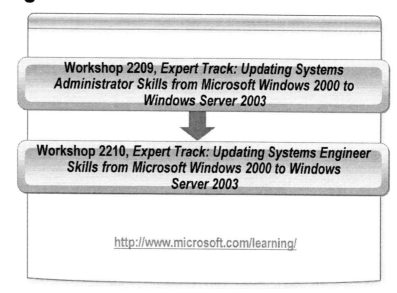

Microsoft Learning develops Official Microsoft Learning Products for computer professionals who design, develop, support, implement, or manage solutions by using Microsoft products and technologies. These learning products provide comprehensive skills-based training in instructor-led and online formats.

Additional Recommended Workshops

Each workshop relates in some way to another workshop. A related workshop may be a prerequisite, a follow-up workshop in a recommended series, or a workshop that offers additional training.

It is recommended that you take the following workshops in this order:

- Workshop 2209, *Expert Track: Updating Systems Administrator Skills from Microsoft Windows 2000 to Windows Server 2003*

- Workshop 2210, *Expert Track: Updating Systems Engineer Skills from Microsoft Windows 2000 to Windows Server 2003*

Other related courses and workshops may become available in the future, so for up-to-date information about recommended courses and workshops, visit the Microsoft Learning Web site.

Microsoft Learning Information

For more information, visit the Microsoft Learning Web site at http://www.microsoft.com/learning/.

Microsoft Certified Professional Program

Exam number and title	Core exam for the following track	Elective exam for the following track
70-296: *Planning, Implementing, and Maintaining a Microsoft Windows Server 2003 Environment for an MCSE Certified on Microsoft Windows 2000*	MCSE on Windows Server 2003	N/A

Microsoft
C E R T I F I E D
Professional

http://www.microsoft.com/learning/

Microsoft Learning offers a variety of certification credentials for developers and IT professionals. The Microsoft Certified Professional program is the leading certification program for validating your experience and skills, keeping you competitive in today's changing business environment.

Related Certification Exam

This workshop helps students to prepare for Exam 70-296: *Planning, Implementing, and Maintaining a Microsoft® Windows Server™ 2003 Environment* for *an MCSE Certified on Microsoft Windows® 2000*.

Exam 70-296 is the core exam for the MCSE on Windows Server 2003 certification. There are no additional core or elective exam requirements for an MCSE on Windows 2000 who passes exam 70-296.

MCP Certifications

The Microsoft Certified Professional program includes the following certifications.

- MCSA on Windows 2000 and MCSA on Windows Server 2003

  The Microsoft Certified Systems Administrator (MCSA) certification is designed for professionals who implement, manage, and troubleshoot existing network and system environments based on the Windows 2000 and Windows Server 2003 platforms. Implementation responsibilities include installing and configuring parts of the systems. Management responsibilities include administering and supporting the systems.

- MCSE on Windows 2000 and MCSE on Windows Server 2003

  The MCSE credential is the premier certification for professionals who analyze the business requirements and design and implement the infrastructure for business solutions based on the Microsoft Windows 2000 and Windows Server 2003 platforms. Implementation responsibilities include installing, configuring, and troubleshooting network systems.

- MCAD

  The Microsoft Certified Application Developer (MCAD) for Microsoft .NET credential is appropriate for professionals who use Microsoft technologies to develop and maintain department-level applications, components, Web or desktop clients, or back-end data services or work in teams developing enterprise applications. The credential covers job tasks ranging from developing to deploying and maintaining these solutions.

- MCSD

  The Microsoft Certified Solution Developer (MCSD) credential is the premier certification for professionals who design and develop leading-edge business solutions with Microsoft development tools, technologies, platforms, and the Microsoft Windows DNA architecture. The types of applications MCSDs can develop include desktop applications and multi-user, Web-based, N-tier, and transaction-based applications. The credential covers job tasks ranging from analyzing business requirements to maintaining solutions.

- MCDBA on Microsoft SQL Server™ 2000

  The Microsoft Certified Database Administrator (MCDBA) credential is the premier certification for professionals who implement and administer Microsoft SQL Server databases. The certification is appropriate for individuals who derive physical database designs, develop logical data models, create physical databases, create data services by using Transact-SQL, manage and maintain databases, configure and manage security, monitor and optimize databases, and install and configure SQL Server.

- MCP

  The Microsoft Certified Professional (MCP) credential is for individuals who have the skills to successfully implement a Microsoft product or technology as part of a business solution in an organization. Hands-on experience with the product is necessary to successfully achieve certification.

- MCT

  Microsoft Certified Trainers (MCTs) demonstrate the instructional and technical skills that qualify them to deliver Official Microsoft Learning Products through Microsoft Certified Technical Education Centers (Microsoft CTECs).

Certification Requirements

The certification requirements differ for each certification category and are specific to the products and job functions addressed by the certification. To become a Microsoft Certified Professional, you must pass rigorous certification exams that provide a valid and reliable measure of technical proficiency and expertise.

 Additional Information See the Microsoft Learning Web site at http://www.microsoft.com/learning/.

You can also send e-mail to mcphelp@microsoft.com if you have specific certification questions.

Acquiring the Skills Tested by an MCP Exam

Official Microsoft Learning Products can help you develop the skills that you need to do your job. They also complement the experience that you gain while working with Microsoft products and technologies. However, no one-to-one correlation exists between Official Microsoft Learning Products and MCP exams. Microsoft does not expect or intend for a course or workshop to be the sole preparation method for passing MCP exams. Practical product knowledge and experience is also necessary to pass the MCP exams.

To help prepare for the MCP exams, use the preparation guides that are available for each exam. Each Exam Preparation Guide contains exam-specific information, such as a list of the topics on which you will be tested. These guides are available on the Microsoft Learning Web site at http://www.microsoft.com/learning/.

Facilities

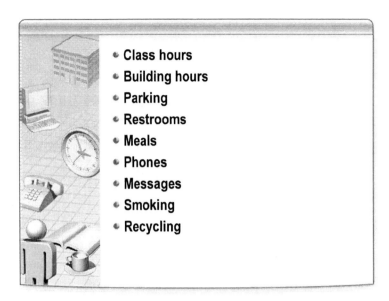

- Class hours
- Building hours
- Parking
- Restrooms
- Meals
- Phones
- Messages
- Smoking
- Recycling

Unit 1: Introduction to Performing Systems Engineer Skills in Windows Server 2003

Contents

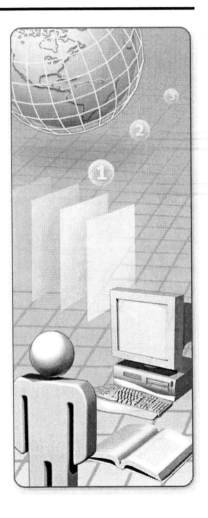

Overview

* **New Features of Windows Server 2003**
* **Classroom Forest and Domain Configuration**
* **Resource Toolkit**
* **Lab: Exploring the Workshop Lab Interface**
* **Lab Discussion**

Microsoft® Windows Server™ 2003 provides new capabilities and tools for the Microsoft Certified Systems Engineer (MCSE). While the systems engineer's job role covers a broad range of tasks, this workshop focuses on the tasks that new features in the operating systems in the Windows Server 2003 family affect, or features that have changed from the Microsoft Windows® Server 2000 family.

This unit introduces these tasks and describes the concepts that the workshop will cover. The lab for this unit introduces you to the scenario used in the workshop and the overall lab environment.

Objectives

After completing this unit, you will be able to:

■ Describe, at a high level, the new features in Windows Server 2003 that pertain to the systems engineer job role.

■ Use the lab environment and locate key resources that are used to complete the labs.

New Features of Windows Server 2003

New features of Windows Server 2003 include:

* DNS namespace design
* Active Directory and replication planning
* DNS and Active Directory implementation
* Troubleshooting tools
* Multiple forests in Active Directory
* Group Policy to deploy and restrict software
* Group Policy advanced security settings
* Routing and Remote Access

In the scenarios for the labs in this course, you are a new employee for Northwind Traders, a fictitious company that has recently acquired a second company, which is the name of the forest on your student computer.

In the labs for each unit, you will perform tasks that focus on the new features of Windows Server 2003 as they pertain to the following topics:

- *Planning a Domain Name System (DNS) namespace design.* Windows Server 2003 provides several enhancements that improve fault tolerance and availability in DNS environments with distributed DNS servers and multiple namespaces. These enhancements include stub zones, conditional forwarding, and application directory partitions for storing DNS information.

- *Planning for Microsoft Active Directory® and replication.* Windows Server 2003 also provides enhancements for Active Directory replication, such as linked value replication (LVR) and Partial Attribute Set (PAS) replication. Most of these enhancements decrease the amount of replication traffic that is sent between domain controllers. Other enhancements, such as universal group membership caching and an improved Inter Site Topology Generator (ISTG), provide additional options for configuring Active Directory sites.

- *Implementing DNS with Active Directory.* By creating stub zones and conditional forwarders, you can make name resolution in your organization more efficient. You can also use application directory partitions to store DNS zone information so that you can control the replication topology, while still making the required DNS records highly available. Windows Server 2003 also provides advanced options for installing domain controllers.

- *Troubleshooting Transmission Control Protocol/Internet Protocol (TCP/IP), name resolution, and Group Policy.* A Windows Server 2003 network in a large organization can be complex. To manage that complexity, Windows Server 2003 provides several new troubleshooting tools to diagnose difficult errors.

- *Planning and implementing multiple forests in Active Directory.* Many of the enhancements in Windows Server 2003 enable deployment of Active Directory in more flexible configurations. These enhancements include forest trusts, which are both easier to configure and more secure than inter-forest trusts in Windows 2000.

- *Using Group Policy to deploy and restrict software.* Windows Server 2003 includes a new option in Group Policy to completely install an assigned software application when a user logs on to the domain. You can also use Group Policy to restrict the software that runs on users' desktops, providing your organization with an additional layer of security against unwanted applications on client workstations.

- *Using Group Policy to set advanced security settings.* Group Policy extends security in Windows Server 2003. You can use Group Policy to manage wireless networks and configure advanced security settings to secure data on computer hard disks.

- *Planning and implementing secure Routing and Remote Access.* Enhancements to this feature include improved remote access policy settings and protocol filtering, and additional options for using network address translation (NAT).

 Important In Windows Server 2003, most services run under dedicated, restricted machine accounts called Local Service and Network Service. These accounts have minimal permissions on the local machine. In Windows 2000, most services ran under Local System Authority, which exposed the system if the service was successfully attacked.

Classroom Forest and Domain Configuration

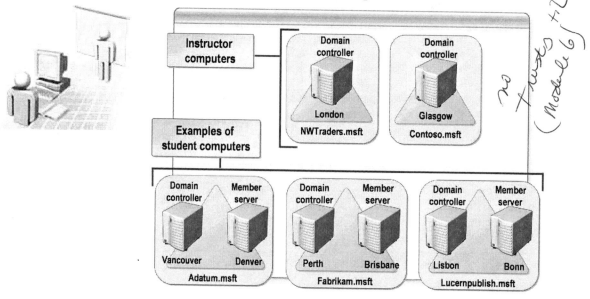

In the labs for this workshop, you are a new employee of Northwind Traders, a company undergoing rapid growth. You have been given systems engineer responsibilities for a newly-acquired subsidiary of Northwind Traders.

The computers in the classroom are configured as multiple forests. There are no trusts created between the forests in the classroom. There are two instructor computers and several student computers.

Instructor Computers

Two instructor computers represent Northwind Traders and its business partner, Contoso Ltd. The first instructor computer, London.NWTraders.msft, is the domain controller in the NWTraders.msft domain. NWTraders.msft is the forest root domain. The second instructor computer, Glasgow.Contoso.msft, is the domain controller in the Contoso.msft domain. Contoso.msft is the forest root domain.

Student Computers

The student computers in the classroom are configured in domains. Each student domain is a forest root domain and includes two servers. One server is a domain controller in the student domain, and the other is a member server in the student domain. For example, one pair of computers is in the Adatum.msft forest. Vancouver (the first student computer) is a domain controller, and Denver (the second student computer) is a member server. The second pair of computers, Perth and Brisbane, is in the Fabrikam.mstft forest, and so on. The table on the lab introduction page lists the server configurations for all of the computers in the lab.

Resource Toolkit

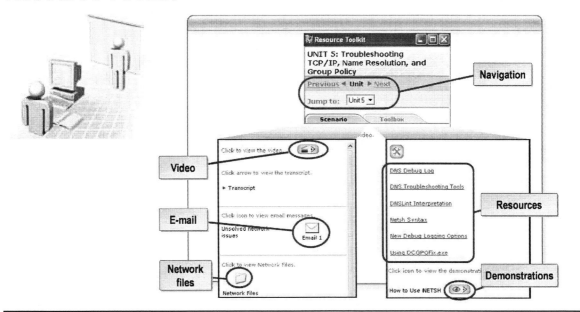

The Resource Toolkit is the user interface for the labs. You navigate to different labs by clicking **Previous** or **Next** or by clicking on the drop-down box. The Toolkit contains two tabs: Scenario and Toolbox.

Scenario Tab

The **Scenario** tab contains the goals and requirements for each lab, including:

- *Video*. Each lab begins with a brief video from your supervisor at Northwind Traders. The videos present the problems that Northwind Traders is facing and the general goal of the lab.

- *E-mail*. E-mail from your supervisor contains additional context and information necessary for completing the lab. The student workbook contains step-by-step tasks that you perform to meet the goals of the e-mail.

- *Network files*. This folder contains additional scenario-based files, such as network diagrams and company policies. You may be prompted in the videos or e-mails to examine these files.

Toolbox Tab

The **Toolbox** tab contains information designed to assist you in performing specific steps in the labs, including:

- *Resources*. Includes facts, tables, practices, and other types of information that you may find useful as you complete the steps in the lab.

- *Demonstrations*. Includes animations that describe processes and demonstrations that show a user clicking through a user interface (UI).

Lab: Exploring the Workshop Lab Interface

In this lab, you will:

* Use the lab interface and locate Toolbox resources
* Create the required Administrator accounts and a regular user account
* Describe the secondary logon service

After completing this lab, you will be able to:

* Use the lab interface and locate Toolbox resources.
* Create the required administrator accounts and a regular user account.
* Describe the secondary logon service.

Estimated time to complete this lab: **30 minutes**

Toolbox Resources

* Using the Workshop Resources
* Guidelines for Creating Strong Passwords
* Secondary Logon Service
* Creating Administrator Accounts

Classroom Domain and Server Configuration

The following table lists the server configuration for all computers in the lab.

Computer name	IP address	Domain name	Role in domain
London	192.168.x.200	Nwtraders.msft	Domain controller for forest root domain
			DHCP server
			DNS server
Glasgow	192.168.x.201	Contoso.msft	Domain controller for forest root domain
Vancouver	192.168.x.1	Adatum.msft	Domain controller for forest root domain
Denver	192.168.x.2	Adatum.msft	Member server in domain
Perth	192.168.x.3	Fabrikam.msft	Domain controller for forest root domain
Brisbane	192.168.x.4	Fabrikam.msft	Member server in domain
Lisbon	192.168.x.5	Lucernpublish.msft	Domain controller for forest root domain
Bonn	192.168.x.6	Lucernpublish.msft	Member server in domain
Lima	192.168.x.7	Litwareinc.msft	Domain controller for forest root domain
Santiago	192.168.x.8	Litwareinc.msft	Member server in domain
Bangalore	192.168.x.9	Tailspintoys.msft	Domain controller for forest root domain
Singapore	192.168.x.10	Tailspintoys.msft	Member server in domain
Casablanca	192.168.x.11	Wingtiptoys.msft	Domain controller for forest root domain
Tunis	192.168.x.12	Wingtiptoys.msft	Member server in domain
Acapulco	192.168.x.13	Thephonecompany.msft	Domain controller for forest root domain
Miami	192.168.x.14	Thephonecompany.msft	Member server in domain
Auckland	192.168.x.15	cpandl.msft (City power and light)	Domain controller for forest root domain
Suva	192.168.x.16	cpandle.msft (City power and light)	Member server in domain
Stockholm	192.168.x.17	Adventureworks.msft	Domain controller for forest root domain
Moscow	192.168.x.18	Adventureworks.msft	Member server in domain
Caracas	192.168.x.19	Blueyonderair.msft	Domain controller for forest root domain
Montevideo	192.168.x.20	Blueyonderair.msft	Member server in domain
Manila	192.168.x.21	Woodgrovebank.msft	Domain controller for forest root domain
Tokyo	192.168.x.22	Woodgrovebank.msft	Member server in domain
Khartoum	192.168.x.23	Treyresearch.msft	Domain controller for forest root domain
Nairobi	192.168.x.24	Treyresearch.msft	Member server in domain

Exercise 1
Exploring the Workshop Lab Interface

In this exercise, you will explore the workshop lab interface and create a non-administrative user account. You will also research the Secondary Logon service in Windows Server 2003.

Tasks	Supporting information
1. Log on using the domain Administrator account and read e-mail.	■ The current server password is **P@ssw0rd**. ■ You can access the Resource Toolkit by using the icon on your desktop. ■ As you work through each step, you will have resources to assist you. For example, if you are not sure what a strong password is, you can access a resource by clicking the **Toolbox** tab. If you already know what to do, you do not need to use the Toolbox resource. See the Toolbox resource, Using the Workshop Resources.
2. Locate the company security policies and read them.	■ Sometimes the same resource will be referenced for a similar task. See the Toolbox resource, Using the Workshop Resources.
3. Create the following administrator accounts for yourself: • An account with permission to modify enterprise settings • An account with permission to modify domain settings • An account with permissions to manage the DNS service	■ Use a name that follows company naming conventions. See the company security policy in the Network Files folder. ■ Set strong passwords for all administrator accounts. ■ You should perform all administrative tasks by using administrator accounts with the minimum required permissions. See the following Toolbox resources: ■ Guidelines for Creating Strong Passwords ■ Creating Administrative Accounts
4. Create a regular user account for yourself.	■ Use a name that follows company naming conventions. See the company security policy in the Network Files folder. ■ Set a strong password for the user account.
5. Respond to the e-mail with a summary of the new features of the Secondary Logon service.	■ To respond to an e-mail, open the e-mail, and then click **Reply**. Type your answer, and then click **Send**. The e-mail will be saved to a folder on your desktop, where you can retrieve it during lab discussion. ■ Some labs require a response via e-mail or a completed worksheet or table. Other labs focus on tasks that you must perform on your computer. See the Toolbox resource, Secondary Logon Service.

Lab E-mail

From: Dale Sleppy

To: Systems Engineers

Sent: Fri Sep 05 13:22:00 2003

Subject: Northwind Traders policies and procedures

I wanted to make sure that you understand a few things. You are in charge of our newly acquired subsidiary. As the lead systems engineer for this forest, your responsibility is to ensure that it interoperates seamlessly with Northwind Traders. Your network consists of 10 Windows Server 2003 servers located in one headquarters office and four branch offices. See the Corporate Locations diagram in the Network Files folder for details.

As you may remember from our interview, security is a top priority for us, and you should always use the most secure method of implementing any task. Failing to comply with this policy can have a serious negative impact on your performance review. Please start by reading the company security policies for acceptable use of administrator and non-administrator accounts and user account naming. They are in the Network Files folder.

Please create the required administrative accounts for yourself. Also, please create a regular user account for yourself, and use that account unless you must be logged on as an administrator. Please set a strong password on the user accounts that you create, as specified in the company security policies.

As you get started, I also want you to inform me about new features of the Secondary Logon service in Windows Server 2003. Please summarize the changes in your reply to this e-mail.

Thanks,

Dale Sleppy

MCSE, MCSA, MCT, CNE, CCNA,

Network +, Server +, Security +, CISSP

Managing Network Engineer

Northwind Traders, Inc.

Lab Discussion

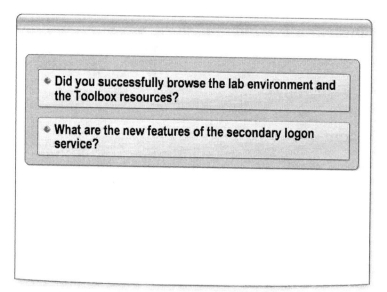

* Did you successfully browse the lab environment and the Toolbox resources?

* What are the new features of the secondary logon service?

Discuss with the class how you implemented the solutions in the preceding lab.

- Were you able to successfully browse the lab environment and the Toolbox resources?
- What are the new features of the secondary logon service?

Unit 2: Planning a DNS Namespace Design

Contents

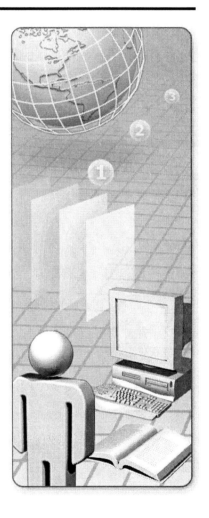

.

Overview

- **Stub Zones and Conditional Forwarding**
- **_MSDCS Availability**
- **Replication Options for Active Directory Integrated DNS Zones**
- **Lab: Planning a DNS Namespace Design**
- **Lab Discussion**
- **Best Practices**

Planning a Domain Name System (DNS) strategy with Microsoft® Windows Server™ 2003 is similar to planning for Microsoft Windows® 2000. However, the following additional features of Windows Server 2003 can considerably improve the namespace design:

- Stub zones

- Conditional forwarding

- DNS and application directory partitions

The purpose of this unit is to enable you to plan a DNS strategy for an enterprise organization by using the new DNS features in Windows Server 2003. The unit focuses on planning issues related to improving fault tolerance in DNS, ensuring DNS resolution across forests, planning for _msdcs zone availability and security to DNS servers in a forest, and creating DNS zones securely and with the least administrative effort.

Objectives

After completing this unit, you will be able to:

- Evaluate existing DNS infrastructure and determine where new Windows Server 2003 features can improve name resolution.

- Determine when to use stub zones versus conditional forwarding.

- Ensure availability of the _msdcs zone.

- Plan Microsoft Active Directory® partitions to replicate zone data when needed.

- Evaluate DNS zone security.

 Additional Information For a more information about planning a DNS namespace strategy and how computers use DNS to locate a domain controller, see the animations *Planning a DNS namespace strategy* and *How Client Computers Use DNS to Locate Domain Controllers and Services*, under **Additional Reading** on the Web page on the Student Materials compact disc.

Stub Zones and Conditional Forwarding

Stub zones	Conditional forwarding
• A read-only copy of a DNS zone that contains only the resource records necessary to identify the authoritative DNS servers for the actual zone	• Configures a DNS server to forward a query it receives to a DNS server depending on the DNS name in the query
• Use when you want a DNS Server hosting a parent zone to remain aware of all the DNS servers authoritative for a child zone	• Use when DNS clients in separate networks should resolve names without querying DNS servers on the Internet
• Send information to all authoritative DNS servers instead of specified DNS servers	• You need to manually configure the conditional forwarding setting on each DNS server that hosts the parent zone

There are some significant improvements in name resolution in Windows Server 2003 DNS over Windows 2000 DNS. The new features in Windows Server 2003 DNS that make DNS administration easier include:

- Stub zones.
- Conditional forwarding.

Stub Zones

A *stub zone* is:

- A read-only copy of a DNS zone that contains only the resource records necessary to identify the authoritative DNS servers for the actual zone.
- Like a secondary zone, except that it replicates only the necessary records of a master zone instead of the entire zone. The stub zone contains only the:
  - SOA (start of authority) resource record.
  - Name Server (NS) resource records.
  - Glue host (A) resource records.

 Important To create a stub zone that is not stored in Active Directory, you need to be a member of the DnsAdmins group. To create a stub zone that is stored in the DomainDnsZones or the ForestDnsZones partition, you need to be a member of the Domain Admins group.

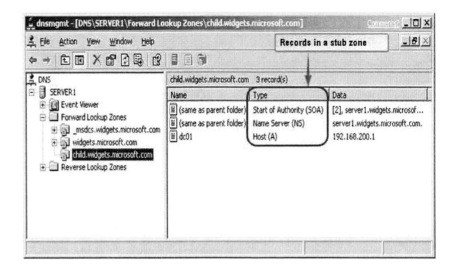

 Note An administrator cannot modify a stub zone's resource records. An administrator must make any desired changes to the resource records in a stub zone in the original primary zone from which the stub zone is derived.

A stub zone keeps a DNS server hosting a parent zone aware of the authoritative DNS servers for its child zone and, thereby, helps it to maintain DNS name resolution efficiency.

 Additional Information For more information about DNS stub zones, see the Support WebCast, "Microsoft Windows Server 2003 DNS: Stub Zones and Conditional Forwarding," at http://support.microsoft.com/default.aspx?scid=/servicedesks/webcasts/wc012103/ wcblurb012103.asp.

Conditional Forwarding

Conditional forwarders are DNS servers used to forward queries according to the domain name contained in the query.

 Important To configure conditional forwarding, you need to be a member of the DnsAdmins group.

Conditional forwarding is used to:

- Forward DNS queries to other DNS servers based on the DNS domain names in the queries. For example, all queries to a DNS server for names ending with nwtraders.msft can be forwarded to a specific DNS server's Internet Protocol (IP) address, or to the IP addresses of multiple DNS servers.

- Manage name resolution among different namespaces in your network. For example, when two companies (Northwind Traders and Contoso) merge or collaborate, they may want to allow clients from the internal namespace of one company to resolve the names of the clients from the internal namespace of another company.

  The administrators at nwtraders.msft may inform the administrators of contoso.msft about the set of DNS servers that they can use for name resolution within nwtraders.msft. In this case, the DNS servers within contoso.msft will be configured to forward all queries for names ending with nwtraders.msft to the designated DNS servers.

Stub Zones vs. Conditional Forwarding

The following table lists the main differences between stub zones and conditional forwarding.

Stub zones	Conditional forwarding
Use when you want a DNS server hosting a parent zone to remain aware of the authoritative DNS servers for one of its child zones.	Use where you want DNS clients in separate networks to resolve each others' names without having to query DNS servers on the Internet. For example, in the case of a company merger, you should configure the DNS servers in each network to forward queries for names in the other network.
Stub zones do not provide the same server-to-server benefit as conditional forwarding. A DNS server hosting a stub zone in one network will reply to queries for names in the other network with a list of all authoritative DNS servers for the zone with that name, instead of the specific DNS servers that you have designated to handle this traffic.	Conditional forwarding is not an efficient method of keeping a DNS server hosting a parent zone aware of the authoritative DNS servers for a child zone. If you use this method, whenever the authoritative DNS servers for the child zone are changed, the conditional forwarder setting on the DNS servers hosting the parent zone would have to be manually configured with the IP address for each new authoritative DNS server for the child zone.

_MSDCS Availability

* **Windows 2000**
  * _msdcs domain is created as part of the parent domain's zone and not as a separate zone
  * Only DNS servers that host the zone for the root domain in the forest contain the _msdcs subdomain for the root domain
* **Windows 2003**
  * When DNS is installed as part of the Active Directory installation process, the Active Directory Installation Wizard creates a separate zone for the _msdcs subdomain and configures it to replicate to all domain controllers in the forest
  * If DNS is not installed and configured during the installation of Active Directory, you must manually create an _msdcs zone and configure it to replicate to all DNS servers in the forest

_msdcs is a subdomain or zone that contains all of the SRV records that domain controllers register when Active Directory is installed. In addition, the _msdcs zone for the root domain in the forest contains all of the SRV records for all of the global catalog servers in the forest.

_MSDCS in Windows 2000

In Windows 2000, the _msdcs domain is not created as a separate zone. It is created as part of the parent domain's zone. Since the _msdcs subdomain is created in DNS as a subdomain of the root domain in the forest, only DNS servers that host the zone for the root domain in the forest contain the _msdcs subdomain for the root domain. When a user logs on, his or her computer must locate a global catalog server by contacting a DNS server that hosts the _msdcs subdomain for the root domain in the forest. If all of the DNS servers that host this zone are in different locations than the user's computer, the logon process takes much longer than if the DNS server is local. To overcome this issue, you can create a secondary copy of the zone for the root domain in each location.

_MSDCS in Windows Server 2003

When a new forest is created on a computer running Windows 2003 Server, and DNS is installed as part of the Active Directory installation process, the Active Directory Installation Wizard creates a separate zone for the _msdcs subdomain of the root domain of the forest. It also configures that zone to replicate to all domain controllers in the forest.

If DNS is not installed and configured on a computer running Windows Server 2003 during the installation of Active Directory, you must manually configure the _msdcs subdomain to be highly available. To make the _msdcs subdomain of the root domain in the forest highly available, create it as a separate zone and configure it to replicate to all DNS servers in the forest. This will make the _msdcs zone available on all Active Directory-integrated DNS servers in the forest.

Replication Options for Active Directory Integrated DNS Zones

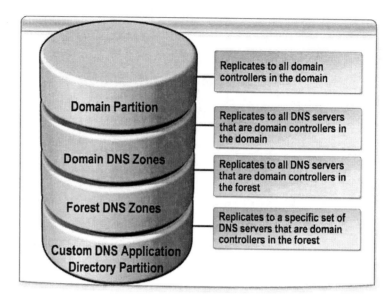

The Active Directory architecture in Windows Server 2003 is very similar to that of Windows 2000, with some added enhancements. In addition to the schema, configuration, and domain partitions included in Windows 2000, Windows Server 2003 also includes an additional partition type, the application directory partition.

Application Directory Partition

Application directory partitions store application-specific information in Active Directory. Each application determines how it will store, categorize, and use application-specific information. Application directory partitions provide Windows Server 2003 with the capability of hosting dynamic data in Active Directory without significantly impacting network performance; it does this by providing the ability to control the scope of replication and placement of replicas. To prevent unnecessary replication of specific application directory partitions, Active Directory administrators can designate which domain controllers in a forest will host specific application directory partitions. The application directory partition is different than a domain partition in that it is not allowed to store security principal objects such as user accounts. In addition, the data in an application directory partition is not stored in the global catalog.

Replication Options for DNS Zones in Active Directory

DNS zones can be stored in the domain or application directory partitions of Active Directory. The following table describes the replication options available for replicating Active Directory integrated DNS zones.

Active Directory Partition	Description
Domain partition	Replicates to all domain controllers in the domain. Windows 2000 DNS servers store Active Directory integrated zones in this partition.
Domain DNS zones	Replicates to all DNS servers that are domain controllers in the domain. This partition is automatically created on Windows Server 2003 domain controllers that are also DNS servers.
Forest DNS zones	Replicates to all DNS servers that are domain controllers in the forest. This partition is automatically created on Windows Server 2003 domain controllers that are also DNS servers.
Custom DNS application directory partition	Replicates to a specific set of DNS servers that are domain controllers in the forest. If you want to replicate a DNS zone only to specific DNS servers within the domain or forest, you can create a new application directory partition for the new zone. The application directory partition must be manually created on each DNS server that will host the zone. Use Ntdsutil.exe to create the new application directory partition and to specify which servers will host a replica of the new partition.

Lab: Planning a DNS Namespace Design

In this lab, you will:

- Determine the most effective DNS method that supports fault tolerance and minimizes bandwidth usage

- Determine which DNS resolution method ensures name resolution among forests

- Plan for _msdcs zone availability and security to DNS servers in a forest

- Create a plan that addresses DNS zone security and uses the least administrative effort

After completing this lab, you will be able to:

- Determine the most effective DNS method that supports fault tolerance and minimizes bandwidth usage.

- Determine which DNS resolution method ensures name resolution among forests.

- Plan for _msdcs zone availability and security to DNS servers in a forest.

- Create a plan that addresses DNS zone security and uses the least administrative effort.

Estimated time to complete this lab: **60 minutes**

Toolbox Resources

If necessary, use one or more of the following resources to help you complete this lab:

- DNS Name Resolution Methods

- What Is a Stub Zone?

- What Is Conditional Forwarding?

- What Are _msdcs Zones?

- _msdcs High Availability

- Active Directory Partitions and Replication Scope

Additional Scenario Information

The DNS network diagram (DNS Configuration.jpg) shows the relationship between your forest and the nwtraders.msft and contoso.msft forests. It includes information about wide area network (WAN) links, Internet connectivity, firewall settings, virtual private networking (VPN) settings, and DNS forwarding configuration. Consider all of this information in your planning process.

In the document, replace:

- *YourForest* with the name of the forest hosted by your student computer.
- *x* with the number of the IP subnet in use in your classroom.
- *y* with the number assigned to your student computer.

Exercise 1
Planning DNS Name Resolution for Your Forest

In this exercise, you will:

- Determine the most effective DNS method that supports fault tolerance and minimizes bandwidth usage.

- Determine which DNS resolution method ensures name resolution among forests.

- Plan for _msdcs zone availability and security to DNS servers in a forest.

- Create a plan that addresses DNS zone security and uses the least administrative effort.

 Important Ensure that you use an account with the lowest level of administrative permissions required for each task in this exercise.

Tasks	Supporting information
1. Determine which DNS resolution method can be best used to ensure fault tolerance, bandwidth, and name resolution between your forest and nwtraders.msft.	■ To implement this requirement, consider the following methods: • Standard forwarding • Secondary zones • Conditional forwarding • Stub zones ■ Plan your implementation, and then fill out the Nwtraders.msft section of the Configuring DNS Name Resolution to Other Forests planning sheet. Be prepared to explain your answer. See the following Toolbox resources: ■ DNS Name Resolution Methods ■ What Is a Stub Zone? ■ What Is Conditional Forwarding? ■ Case Study: Using Stub Zones for Fault Tolerance ■ Case Study: Using Conditional Forwarders to Reduce WAN Bandwidth for DNS Name Resolution

(continued)

Tasks	Supporting information
2. Determine which DNS resolution method can be best used to ensure fault tolerance, bandwidth, and name resolution between your forest and contoso.msft.	▪ To implement this requirement, consider the following methods: • Standard forwarding • ~~Secondary zones~~ • ~~Conditional forwarding~~ *Yes.* • Stub zones ▪ Plan your implementation, and then fill out the Contoso.msft section of the Configuring DNS Name Resolution to Other Forests planning sheet. Be prepared to explain your answer. See the following Toolbox resources: ▪ DNS Name Resolution Methods ▪ What Is a Stub Zone? ▪ What Is Conditional Forwarding? ▪ Case Study: Using Stub Zones for Fault Tolerance ▪ Case Study: Using Conditional Forwarders to Reduce WAN Bandwidth for DNS Name Resolution
3. Plan for _msdcs zone availability and security to all DNS servers in your forest.	▪ To implement this requirement, consider the following issues: • Zone type • Security for updates and zone transfers *AD Integrated* • Active Directory integration • Replication scope ▪ Plan your implementation, and then fill out the _msdcs High Availability planning sheet. Be prepared to explain your answer. *Allow transfer primary* See the following Toolbox resources: ▪ What Are _MSDCS Zones? ▪ _MSDCS High Availability ▪ Active Directory Partitions and Replication Scope
4. Plan the creation of new zones. When DNS zones are created, you want to ensure that they are as secure as possible and available to all DNS servers with the least administrative effort.	▪ To implement this requirement, consider the following issues: • Zone type *primary* • Security for updates and zone transfers ✓ • Active Directory integration ✓ • Replication scope ✓ ▪ Plan your implementation, and then fill out the New Zone planning sheet. Be prepared to explain your answer. See the Toolbox resource Active Directory Partitions and Replication Scope.

(handwritten: _msdcs.woodgrovebank.msft)

Configuring DNS Name Resolution to Other Forests Planning Sheet

Forest name	Name resolution method	Firewall settings	Justification
Nwtraders.msft	2ndary zone / 192.168.1.200	Conditional	limited bandwidth
Contoso.msft	Conditional / .201		

_MSDCS High Availability Planning Sheet

Zone type and replication scope	Security	Justification
Forwarder	High	

New Zone Planning Sheet

Zone type and replication scope	Security	Justification

Lab E-mail

From: Dale Sleppy

To: Systems Engineers

Sent: Fri Sep 05 13:26:58 2003

Subject: Company Acquisitions and DNS Issues

As the new systems engineer for our recently acquired company, I wanted to make you aware of our company acquisition strategy. Please review the DNS Network diagram in the Network Files folder to familiarize yourself with our structure. In anticipation of our future plans, I would like to let you know of some issues that our IT staff has brought to my attention:

- Currently, name resolution between your forest and nwtraders.msft uses standard forwarding. There is concern that this is not as fault tolerant or as flexible as we can be. We need to explore other possibilities. Bandwidth between the two forests is always of concern. It is anticipated that there will be a moderate number of DNS queries between your forest and the nwtraders.msft forest.controller for the same domain or no global catalog for the same forest.

- Our company plans to acquire Contoso Ltd. After the acquisition, we will keep the domain name of contoso.msft. We want to have name resolution in place before we move forward and establish a trust. It is anticipated that there will only be a limited number of DNS queries between your forest and the contoso.msft forest on an ongoing basis.

- In the past, we have received serious replication and logon errors due to the unavailability of the _MSDCS in the forest root. You need to ensure that this is highly available in your forest as we grow to prevent problems that we have experienced in the past.

- Whenever zones are created in DNS, we need to ensure that they are secure and highly available to all DNS servers in the forest. Whenever possible, use methods that will automate the creation of these zones.

The firewalls that are set up between the companies are maintained by different security teams. If you need to change any of the firewall settings, please justify your decisions in your plan.

Please submit the DNS planning worksheet with your DNS vision and recommendations based on the aforementioned requirements.

Dale Sleppy

MCSE, MCSA, MCT, CNE, CCNA,

Network +, Server +, Security +, CISSP

Managing Network Engineer

Northwind Traders, Inc.

Lab Discussion

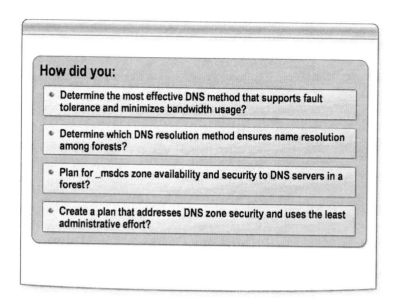

How did you:

- Determine the most effective DNS method that supports fault tolerance and minimizes bandwidth usage?

- Determine which DNS resolution method ensures name resolution among forests?

- Plan for _msdcs zone availability and security to DNS servers in a forest?

- Create a plan that addresses DNS zone security and uses the least administrative effort?

Discuss with the class how you implemented the solutions in the preceding lab.

- What account or accounts did you use to complete the tasks in the lab?

- Which DNS name resolution method did you use to resolve names between your forest and the nwtraders.msft forest to meet the fault tolerance and bandwidth requirements? Did you:
  - Use standard forwarding?
  - Use secondary zones?
  - Use stub zones?
  - Use conditional forwarding?
  - Use a different solution?
  - Specify any security restrictions for your choices?

- Which DNS name resolution method did you use to resolve names between your forest and the contoso.msft forest? Did you:
  - Use standard forwarding?
  - Use secondary zones?
  - Use stub zones?
  - Use conditional forwarding?
  - Use a different solution?
  - Specify any security restrictions for your choices?

- How did your plan allow for _msdcs availability? Did you:
  - Specify creation of a primary or secondary zone?
  - Specify that the zone would be stored in a zone database file, or be Active Directory integrated?
  - Specify a particular Active Directory partition in which to store the zone data?
  - Specify any security configurations settings for the data in this zone?
- How did your plan allow for creating new DNS zones? Did you:
  - Specify manual creation of each zone on each DNS server?
  - Specify that the zone would be stored in a zone database file, or be Active Directory integrated?
  - Specify a particular Active Directory partition in which to store the zone data?
  - Specify any special security configuration options?

Best Practices

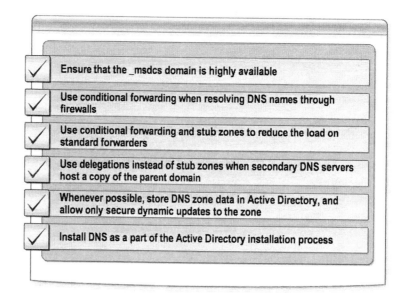

✓	Ensure that the _msdcs domain is highly available
✓	Use conditional forwarding when resolving DNS names through firewalls
✓	Use conditional forwarding and stub zones to reduce the load on standard forwarders
✓	Use delegations instead of stub zones when secondary DNS servers host a copy of the parent domain
✓	Whenever possible, store DNS zone data in Active Directory, and allow only secure dynamic updates to the zone
✓	Install DNS as a part of the Active Directory installation process

- *Ensure that the _msdcs domain is highly available.* Create a new zone for the _msdcs subdomain of the root domain in the forest, and store it in the ForestDNSZones application directory partition so that it will replicate to all DNS servers in the forest.

- *Use conditional forwarding when resolving DNS names through firewalls.*

- *Use conditional forwarding and stub zones to reduce the load on standard forwarders.* To reduce the load on a standard forwarder, use stub zones or conditional forwarding for resolving disjointed name spaces within your network. Be cautious when using stub zones to replace zone delegations, because unlike delegation records, stub zones are not replicated to servers that host secondary copies of the parent DNS zone.

- *Whenever possible, store DNS zone data in Active Directory, and allow only secure dynamic updates to the zone.*

- *Install DNS as a part of the Active Directory installation process.* Installing DNS as part of the Active Directory installation process ensures that DNS is correctly configured to support Active Directory. In addition, if you use this process on the first DNS server in the forest, the _msdcs zone is automatically created and configured to replicate to all DNS servers in the forest.

Unit 3: Planning Active Directory Deployment

Contents

Overview

- New Replication and Active Directory Features
- Linked Value Replication
- Partial Application Set Replication
- Lab: Planning Active Directory Deployment
- Lab Discussion
- Best Practices

Microsoft® Windows Server™ 2003 includes a number of improvements that enable more efficient replication of data within a forest and between sites. In this unit, you will learn how to use the new features to plan and optimize a Microsoft Active Directory® deployment.

Objectives

After completing this unit, you will be able to:

- Evaluate the placement of global catalog servers.
- Plan optimal replication by exploiting the following replication enhancements:
  - Linked value replication
  - Partial attribute set (PAS) replication
  - Inter-Site Topology Generator (ISTG) improvements
- Evaluate forest and domain functionality (versioning) levels.

New Replication and Active Directory Features

* Universal group membership caching
* Partial attribute set replication
* Linked value replication
* Replica domain controller deployment
* New Net Logon service and DNS settings
* Inter-Site Topology Generator enhancements

The Active Directory replication features of Windows Server 2003 are very similar to those of Microsoft Windows® 2000, but there are some new enhancements and efficiencies for network administrators and engineers.

A few of the new Windows Server 2003 features are:

- *Universal group membership caching.* After a user logs on for the first time, the domain controller at that site can obtain the user's universal group membership data from a global catalog server. The domain controller will cache it locally and periodically refresh it so that users can log on without having a global catalog server at that site. However, other applications that depend upon the global catalog cannot benefit from universal group membership caching. For example, applications like Exchange™ 2000 and Exchange Titanium™ in remote offices cannot take advantage of universal group membership caching and must still have access to a global catalog.

- *Partial attribute set replication.* When the PAS is modified, full resynchronization of the global catalog servers in the forest does not occur. Instead, synchronization only includes the attributes that have changed.

- *Linked value replication (LVR).* Groups are not replicated at the object level. Instead, group *membership* is replicated.

- *Replica domain controller deployment.* You can create an additional domain controller by using a system state backup from a domain controller running Windows Server 2003. This ensures minimal impact on wide area network (WAN) links during the installation of domain controllers in remote sites.

- *New Net Logon service and Domain Name System (DNS) settings.* This feature ensures that domain controllers register only needed service (SRV) resource records.

- *Inter-Site Topology Generator enhancements.* The ISTG now uses improved algorithms and can scale to support forests with a greater number of sites than Windows 2000 can. Because all domain controllers in the forest running the ISTG role must agree on the inter-site replication topology, the new algorithms are not activated until the forest is at the Active Directory forest functionality level of Windows Server 2003.

Important To install a replica domain controller, you must be a member of the Domain Admin group.

 Important To enable Universal Group Caching, you must be a member of the Domain Admin group (In the forest root domain) or be a member of Enterprise Admins, or have been delegated the appropriate permissions.

 Note Some of the preceding features are not available until the forest or domain is at the required level to support the feature.

Do you need logging in a
GC to only Admin
No - domain Admin

native mode domain -
Kerbose Token looks for
access user SID
session ticken

Univ. Groups only stored on GC
not domain controller

forest func level needs to be 2003

mixed
native
2003 to can raise forest func levels =
LVR

DCpromo ✓ ✓ ? ⊗
dcpromo /advanced
Ntsidit

Linked Value Replication

* Only the group member that has changed gets replicated
* LVR eliminates the possibility of overwriting changes to the group that another administrator makes
* The forest functional level must be raised to Windows Server 2003

In Windows 2000, when you change the information for a member of a group, the entire group gets replicated. This might cause changes that another administrator makes to be unintentionally overwritten.

In Windows Server 2003, when you change a member of a group, the new LVR functionality replicates only the information that you changed.

 Note To use LVR, the forest functional level must be at Windows Server 2003 or Windows Server 2003 interim.

 Important To raise the forest functional level, you must be a member of the Enterprise Admins group.

 Animation The animation *Linked Value Replication in Windows Server 2003* is located in the **Media** folder on the Web page on the Student Materials CD.

Partial Application Set Replication

* **Windows 2000:** When a newly added attribute is replicated, all object attributes in the global catalog get synchronized
* **Windows Server 2003:** When an attribute is added, only the new attribute gets replicated

Active Directory defines a base set of attributes for each object in the directory. Each object and some of its attributes (such as universal group memberships) are stored in the global catalog. If you use the Active Directory Schema snap-in, you can specify additional attributes to be kept in the global catalog.

■ In Windows 2000 forests, extending the PAS causes a full synchronization of all object attributes stored in the global catalog for all domains in the forest. In a large multi-domain forest, this synchronization can cause significant network traffic.

■ Between domain controllers enabled as global catalogs that are running Windows Server 2003, only the newly added attribute gets replicated.

Lab: Planning Active Directory Deployment

In this lab, you will plan for:

- Remote site global catalogs
- Reducing global catalog replication traffic
- Reducing replication traffic
- Deployment of branch office domain controllers
- Remote office domain controllers
- Site expansion

Objectives

After completing this lab, you will be able to plan for:

- Remote site global catalogs.
- Reducing global catalog replication traffic.
- Reducing Active Directory replication traffic.
- Deployment of branch office domain controllers.
- Remote office domain controllers.
- Site expansion.

Estimated time to complete this lab: **45 minutes**

Toolbox Resources

If necessary, use one or more of the following resources to help you complete this lab:

- What Is Universal Group Membership Caching?
- Enabling Universal Group Membership Caching
- Windows 2000 and Windows Server 2003 Universal Group Availability Solutions
- PAS Replication Comparison Between Windows 2000 and Windows Server 2003
- What Is a Partial Attribute Set?
- Demonstration of Linked Value Replication
- What Is Linked Value Replication?
- Optimizing Remote Domain Controller Installation over WAN Links
- Automated Site Coverage
- Inter-Site Topology Generator

Exercise 1
Planning Active Directory Deployment

In this exercise, you will investigate the issues raised in your manager's e-mail message and create a plan that exploits new features of Windows Server 2003 to resolve those issues.

The following table lists specific questions to answer in your e-mail. When you need additional information, use your Toolbox resources.

 Important Ensure that you use an account with the lowest level of administrative permissions required for each task in this exercise.

Questions	Supporting information
1. Determine the best method for enabling users at a remote site to log on when a WAN link is not available. *[handwritten: Active D.R. step & service / Enable UGM Caching / or make every / domain controller / a global catalog / GC]*	▪ To enable users to reliably log on when WAN links are not available, you could: • Modify the registry to remove the global catalog requirement. • Configure a local domain controller as a global catalog server. • Enable universal group membership caching. ▪ Decide which approach is the best solution, and be prepared to explain your answer. See the following Toolbox resources: ▪ What Is Universal Group Membership Caching? ▪ Enabling Universal Group Membership Caching ▪ Windows 2000 and Windows Server 2003 Universal Group Availability Solutions
2. Evaluate features that will prevent global catalog servers from forcing a full replication of the PAS when modified, and explore ways to prevent saturation of WAN links that connect global catalog servers when schema modifications occur.	See the Toolbox resource, What Is a Partial Attribute Set? *[handwritten notes]*
3. Determine which Windows Server 2003 features will allow groups with greater than 5000 members, and evaluate group membership data conflicts.	See the following Toolbox resources: ▪ What Is Linked Value Replication? ▪ Features Available in Domain and Forest Functional Levels *[handwritten: need to migrate SID history]*
4. Examine ways to deploy a replica domain controller in a branch office without seriously impacting WAN links.	See the Toolbox resource, Optimizing Remote Domain Controller Installation over WAN Links. *[handwritten: depromo / adv]*

(*continued*)

Questions	Supporting information
5. Determine how to ensure that users only log on to domain controllers at their sites or the main site.	See the Toolbox resource, Automated Site Coverage. *Net logon Service*
6. Plan for site expansion to incorporate multiple sites.	▪ The Windows 2000 ISTG has difficulty replicating a large number of sites. ▪ The recommended practice with Windows 2000 is to manually create the site topology if there are over 100 sites. See the Toolbox resource, Inter-Site Topology Generator.

features avail in Dom'/ Forest Func levels:
can rename domain controllers
W 2003 domain func level does not matter (errw
(e.g. if you merge w/ ~~domain~~ company)
painful.

Lab E-mail

From: Dale Sleppy

To: Systems Engineers

Sent: Fri Sep 05 14:00:38 2003

Subject: Branch Office Considerations

I want to familiarize you with some of the problems that we encountered in the main office with a deployment of Windows 2000 to a branch office before you were hired. To give you a better understanding of the network topology of your forest, I have put a diagram of the current network topology in the Network Files folder.

We have had a lot of issues with previous branch office deployments pertaining to domain controller installations and replication. In the following list, I have outlined some of the major issues that we encountered in our deployments. Please consider these when you plan your design:

- When WAN links have failed in the past, some users in our branch offices could not log on to the network, even though there was a domain controller at their site to authenticate them. You might encounter this in the Ashland site, because that link has failed in the past. Unfortunately, due to Ashland's location, that is the best link we can get.

- When we added new objects to be replicated to the global catalog servers, the WAN links to the global catalog servers became saturated and unusable for hours.

- There have been problems when multiple organizational unit (OU) administrators added members to the same groups, and some of their changes were lost. We have also had problems adding more than 5000 users to a group.

- When we installed domain controllers in branch offices, the installation process took over eight hours to replicate the Active Directory database and seriously reduced WAN link availability.

- Users in branch office remote sites were authenticating against domain controllers at other sites that were connected by slow WAN links. This caused some of the slowest WAN links to be taxed by authentication traffic from other sites.

- We plan to expand your forest over the next few years to include over 300 new branch office locations. Please let me know if there are any replication concerns that I should be aware of.

Think about what we can do to prevent these issues from happening to your forest, and reply in an e mail message outlining possible solutions or Windows Server 2003 features that we can use to resolve these issues.

Dale Sleppy

MCSE, MCSA, MCT, CNE, CCNA,

Network +, Server +, Security +, CISSP

Managing Network Engineer

Northwind Traders, Inc.

Lab Discussion

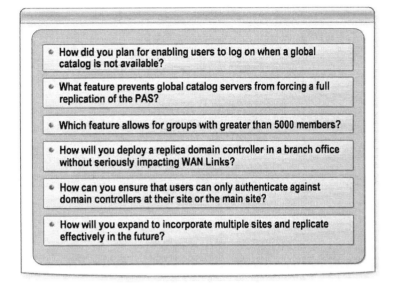

Discuss with the class how you implemented the solutions in the preceding lab.

- What account or accounts did you use to complete the tasks in the lab?
- How did you plan for enabling users to log on when a global catalog is not available?
  - Would you make the site domain controller a global catalog?
  - Would you disable the global catalog requirement in the registry for the local computers?
  - Would you use Windows Server 2003 universal group membership caching?
- What feature prevents global catalog servers from forcing a full replication of the global catalog?
  - Do you need to be at a certain forest or domain functionality level?
  - What are the effects on computers running Windows 2000?
- Which 2003 feature allows groups with greater than 5000 members?
  - Would you create multiple groups and use group nesting?
  - Would you use LVR?
  - What are the advantages of using LVR?
- How do you plan to deploy a replica domain controller in a branch office without seriously affecting WAN links?
  - Should you install locally and send the domain controller to the remote site?
  - Should you use the advanced Dcpromo features?
  - Can advanced Dcpromo be used for global catalog installation?

- How can you ensure that users can only authenticate against domain controllers in their sites or the main site?

  - How can you ensure that the branch office domain controllers register SRV records for only their sites?

  - What are the advantages?

  - What are the disadvantages?

- How do you plan to expand to incorporate multiple sites and replicate effectively?

  - Are there ISTG issues that will limit the number of sites on a domain?

  - Should you create the site topology manually?

Best Practices

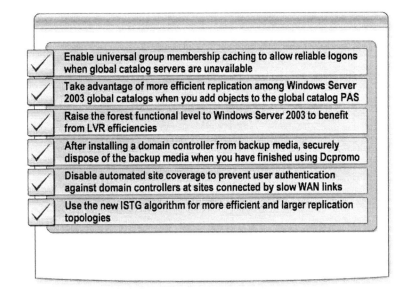

- *Enable universal group membership caching to allow reliable logons when global catalog servers are unavailable.* After universal group caching is enabled, the domain controller caches group membership data locally and periodically refreshes it so that users can log on without requiring a global catalog server at that site.

- *Take advantage of more efficient replication among Windows Server 2003 global catalogs when you add objects to the global catalog PAS.* Only Windows Server 2003 servers can recognize incremental PAS replication.

- *Raise the forest functionality level to Windows Server 2003 to benefit from LVR efficiencies.* The Windows Server 2003 forest functionality level provides the most efficient replication. It can leverage LVR, the improved ISTG algorithm, and PAS replication.

- *After installing a domain controller from backup media, securely dispose of the backup media when you have finished using Dcpromo.* Compromised media could lead to the disclosure of sensitive data stored in Active Directory, such as user accounts.

- *Use automated site coverage to prevent user authentication against domain controllers at sites connected by slow WAN links.* This will make the most efficient use of WAN links, because authentication traffic can be controlled to sites with the fastest WAN links.

- *Use the new ISTG algorithm for more efficient and larger replication topologies.* This will use less overhead for calculations and allow for larger site topologies.

Unit 4: Implementing DNS with Active Directory

Contents

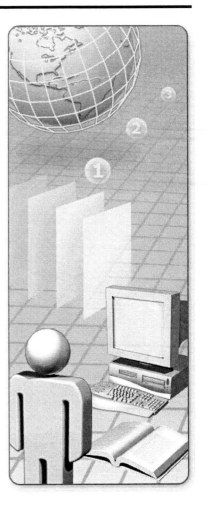

Microsoft

Overview

- **Application Directory Partitions**
- **Lab: Implementing DNS with Active Directory**
- **Lab Discussion**
- **Best Practices**

Microsoft® Windows Server™ 2003 provides several enhancements to Microsoft Active Directory® and Domain Name System (DNS) that increase flexibility and scalability, including:

- Advanced options for Active Directory installation that enable the initial population of the Active Directory database from a restored backup of another domain controller.

- New options for configuring zone delegation and forwarding in DNS, and new ways to ensure availability of DNS information to clients.

- New tools for managing and troubleshooting DNS issues.

- Application directory partitions that can store information in Active Directory and control the replication topology for the partition.

Some enhancements are limited to Windows Server 2003 servers and may not be available in a mixed network comprised of Windows Server 2003 and earlier operating systems. To use these features, you may need to raise the functional level of the Active Directory domain and forest.

Objectives

After completing this unit, you will be able to:

- Install Active Directory by using the advanced features of the Active Directory Installation Wizard.

- Install and configure DNS.

- Implement a conditional forwarder.

- Create stub zones.

- Ensure high availability on the _MSDCS subdomain.

- Create a DNS forward lookup zone.

- Raise domain and forest functionality.

- Create a new application directory partition.

- Set the replication scope of a new application directory partition.

Application Directory Partitions

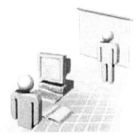

Application directory partitions:
* **Require DNS names**
* **Can be managed by using:**
  * NTDSUtil
  * DNSCmd for DNS application directory partitions
* **Require a security descriptor reference domain**

Application directory partitions store application-specific information in Active Directory and are new in Windows Server 2003. Each application determines how it will store, categorize, and use application-specific information. An application partition is different than a domain partition in that it is not allowed to store security principal objects such as user accounts. In addition, the data in an application partition is not stored in the global catalog. To prevent unnecessary replication of specific application partitions, Active Directory administrators can designate which domain controllers in a forest host specific application partitions.

DNS in Windows Server 2003 is configured by default to use application directory partitions. Windows Server 2003 creates the ForestDNSZones and DomainDNSZones partitions automatically when you install DNS during the Active Directory installation process, or if you install Active Directory on a server running the DNS service. You can also create a new application partition to store DNS information.

 Important To create or modify active directory partitions, you need to be a member of the Domain Admins group, or the Enterprise Admins group.

DNS Names

Application directory partitions require DNS names. They can be:

- Child domains to domain directory partitions, for example,
  Dc=AppPartition,dc=fabrikam,dc=msft

- An additional tree in the Active Directory forest, for example, dc=AppPartition

- A child of another application directory partition, for example,
  Dc=AppPartition1,dc=AppPartition2

Managing Application Directory Partitions

You can manage DNS application directory partitions by using DNSCmd. You can manage all application directory partitions by using the NTDSUtil command line tool. The **Domain Management** menu contains the commands used to manage application directory applications.

With NTDSUtil, you can:

- Add a partition, for example, Add NC DC=AppPartition London.nwtraders.msft

- Delete a partition, for example, Delete NC DC=AppPartition

- Configure replication topology for partitions, for example, Add NC Replica DC=AppPartition London2.nwtraders.msft

 Note One of the benefits of application directory partitions is that replicas for the partitions can be configured on any domain controller in the forest. This means that you can use the partition to replicate information between domains in a forest.

Security Descriptor Reference Domain

Because application directory partitions cannot contain security principals, you must assign permissions to the objects in the partition from a domain in the forest, known as a *security descriptor reference domain*. Each application directory partition is installed with a security descriptor reference domain, and default permissions to the partition objects are granted to members of that domain. If you create an application directory partition as a child partition of an existing domain partition, the domain partition becomes the security descriptor reference domain. If you create an application directory partition as a *new* tree in the forest, the forest root domain is used.

 Note If you demote or remove the domain controller that contains the last replica of an application directory partition from the network, the application partition will be deleted. If you must retain the partition, ensure that it is replicated to another domain controller before demoting the domain controller.

Lab: Implementing DNS with Active Directory

In this lab, you will:

- Install Active Directory by using the advanced features of the Active Directory Installation Wizard
- Install and configure DNS
- Implement a conditional forwarder
- Create stub zones
- Ensure high availability on the _MSDCS subdomain
- Create a DNS forward lookup zone
- Raise domain and forest functionality
- Create a new application directory partition
- Set the replication scope of a new application partition

After completing this lab, you will be able to:

- Install Active Directory by using the advanced features of the Active Directory Installation Wizard.

- Install and configure DNS.

- Implement a conditional forwarder.

- Create stub zones.

- Ensure high availability on the _MSDCS subdomain.

- Create a DNS forward lookup zone.

- Raise domain and forest functionality.

- Create a new application directory partition.

- Set the replication scope of a new application partition.

Estimated time to complete this lab: **180 minutes**

Toolbox Resources

- Features of the /adv Switch
- Configuring Conditional Forwarding
- Features of Ping, DNSLint, and NSLookup
- Using DNSLint
- Active Directory Partitions and Replication Scope
- DNS Security Features in Windows Server 2003
- Creating a DNS Stub Zone
- _MSDCS Zones
- Manually Configuring _MSDCS Zones
- Detailed Steps: Manually Configuring _MSDCS Zones
- How to Determine Successful Registration of _MSDCS Records
- Using Repadmin /syncall /P to Force DNS Replication
- Raising Domain Level Functionality
- Raising Forest Level Functionality
- Features Available in Forest and Domain Functional Levels
- Verifying Functional Level by Using LDP
- Verifying Functional Level by Using ADSIEdit
- Creating a New Application Directory Partition by Using NTDSUtil
- Setting Replication Scope for a New Application Directory Partition by Using NTDSutil
- Using DNScmd to Verify Creation of Application Partitions

Exercise 1
Implementing Active Directory and DNS

In this exercise, an e-mail directs you to implement Active Directory and DNS and lists several conditions that you must fulfill as you perform the exercise.

 Important To install an additional domain controller in forest, you must be a member of the Domain Admins group. To manage the DNS configuration, you must be a member of the DnsAdmins group. To add DNS zones to the domain partition or the DomainDNSZones, you must be a member of the Domain Admins group. To add DNS zones to the ForestDNSZones, you must be a member of the Enterprise Admins group.

 Important Ensure that you use an account with the lowest level of administrative permissions required for each task in this exercise.

Backup 1st.

Tasks	Supporting information
1. Install Active Directory on branch office member server by using the advanced features of the Active Directory Installation Wizard. The new domain controller should not be a global catalog server.	▪ The restored System State files must be accessible to the member controller on a local drive. Plan your restore of the System State data accordingly. See the Toolbox resource, Features of the /adv Switch. *default first st name AD* *Monkeyed w/ Sites & Services. Admin Pool*
2. Move the promoted domain controller from the Domain Controllers organizational unit to the child organizational unit under it called Branch Office.	▪ The second domain controller installed in your domain is the branch office domain controller. *Active Dir Domains & trusts*
3. Configure the domain DNS zone so that it is stored in Active Directory and replicated to all domain controllers that are also DNS servers in your domain.	▪ This step must be completed only on the first domain controller installed in your domain. ▪ The domain DNS zone should be available on all domain controllers that are DNS servers in your domain. This zone should not be replicated to domain controllers in other domains in the forest. See the Toolbox resource, Active Directory Partitions and Replication Scope. *DNS*
4. Install DNS on the branch office domain controller. Confirm that the domain DNS zone is replicated to the second domain controller. Configure the branch office domain controller to use the local DNS server.	▪ To configure the Active Directory DNS zone so that it is replicated to all domain controllers that are also DNS servers, add it to the DomainDNSZones. See the following Toolbox resources: ▪ Using Repadmin /syncall /P to Force DNS Replication ▪ Active Directory Partitions and Replication Scope

(continued)

Tasks	Supporting information

5. Ensure that DNS name resolution within the domain is successful.	■ To ensure that name resolution is working, you could use:
	  • Ping.
	  • DNSLint.
	  • NSLookup.
	■ Decide which approach is the best solution, and be prepared to explain your answer.
	■ To ensure that the servers are not using a cached copy of a DNS resolution, open a command prompt, and type **IPConfig /flushdns**
	■ DNSLint is included in the Windows Server 2003 Support Tools. The installation files for the Support Tools are located in the Support folder on the server installation compact disc or in the \\London\Setup shared folder.
	
	See the following Toolbox resources:
	  ■ Features of Ping, DNSLint, and NSLookup
	  ■ Using DNSLint

6. Implement a conditional forwarder from both the headquarters office DNS server and the branch office DNS server for the contoso.msft domain.	■ Conditional forwarding must be configured on each individual DNS server. ■ Glasgow.Contoso.msft (192.168.*x*.201) is the primary name server for the Contoso.msft domain. See the Toolbox resource, Configuring Conditional Forwarding.

7. Ensure that conditional forwarding works.	■ To ensure that the conditional forwarding is working, you could use:
	  • Ping.
	  • DNSLint.
	  • NSLookup.
	■ Decide which approach is the best solution, and be prepared to explain your answer.
	
	See the following Toolbox resources:
	  ■ Features of Ping, DNSLint, and NSLookup
	  ■ Using DNSLint

(continued)

Tasks	Supporting information
8. Ensure high availability of the _MSDCS subdomain.	■ Determine how you will configure the _MSDCS zone so that it will be highly available throughout your Active Directory forest while still minimizing the zone transfer traffic for the zone. To implement this requirement, you could:
	• Configure the _MSDCS zone so that it is stored in the ForestDNSZones partition.
	• Create a new application directory partition, and store the _MSDCS zone information in that partition.
	• Configure the _MSDCS zone so that it is stored in the DomainDNSZone partition.
	■ Decide which approach is the best solution, and be prepared to explain your answer.
	■ On one of the domain controllers, configure the _MSDCS zone to match your decision.
	See the following Toolbox resources:
	■ MSDCS Zones
	■ Manually Configuring _MSDCS Zones
	■ Detailed Steps: Manually Configuring _MSDCS Zones
	■ Active Directory Partitions and Replication Scope
9. Verify that all of the required resource records are available in the _MSDCS zone after the configuration change.	■ To check the entries in the _MSDCS zone, you can use:
	• The DNS administration tool.
	• NSLookup, querying just the contents of the _MSDCS zone.
	■ Decide which approach is the best solution, and be prepared to explain your answer.
	See the Toolbox resource, How to Determine Successful Registration of _MSDCS Records.
10. Create a new DNS forward lookup zone for fourthcoffee.msft on the Headquarters DNS server.	■ Ensure that all DNS servers in your domain receive zone information for the new zone.
	See the Toolbox resource, Active Directory Partitions and Replication Scope.

(continued)

Tasks	Supporting information
11. Create a new DNS forward lookup zone for wingtiptoys.msft on the Branch Office DNS server.	■ Ensure that all DNS servers in your domain receive zone information for the new zone. See the Toolbox resource, Active Directory Partitions and Replication Scope.
12. Verify that the zones were automatically created on the reciprocal (other) servers.	■ Based on your decision on how to configure the new DNS zones, ensure that the DNS zone is replicated to all required DNS servers. See the Toolbox resource, Using Repadmin /syncall /P to Force DNS Replication.

Exercise 2
Creating Application Directory Partitions

In this exercise, an e-mail directs you to create application directory partitions and lists several conditions that you must fulfill as you perform the exercise. You will also configure a DNS stub zone to use the application directory partition.

 Important Ensure that you use an account with the lowest level of administrative permissions required for each task in this exercise.

Tasks	Supporting information

1. Raise domain and forest functionality to needed levels.	▪ Based on the company requirements and the types of domain controllers in operation, decide what functional level is most appropriate for your organization. To meet this requirement, you could:
	  • Leave the domain and forest at the current levels.
	  • Set the domain and forest levels at the Microsoft Windows® 2000 Native functional level.
	  • Set the domain and forest levels at the Windows Server 2003 functional level.
	
	▪ Decide which approach is the best solution, and be prepared to explain your answer.
	
	See the following Toolbox resources:
	  ▪ Features Available in Forest and Domain Functional Levels
	  ▪ Raising Domain Level Functionality
	  ▪ Raising Forest Level Functionality

2. Verify the domain and forest functional levels.	▪ To verify functional levels, you can use:
	  • Active Directory Domains and Trusts.
	  • ADSIEdit.msc.
	  • LDP.exe.
	▪ Decide which approach is the best solution, and be prepared to explain your answer.
	
	See the following Toolbox resources:
	  ▪ Verifying Functional Level by Using ADSIEdit
	  ▪ Verifying Functional Level by Using LDP

3. Create a new application directory partition named AppPartition.local.	▪ Perform this action on one of the domain controllers in your domain. See the Toolbox resource, Creating a New Application Directory Partition by Using NTDSUtil.

(continued)

Tasks	Supporting information
4. Set replication scope of new application partition so that it is replicated to both the Main Office and Branch Office DNS servers.	▪ Use NTDSUtil to add a new replica for an application directory partition. Add the partition to the server that does not have a copy of the partition. 🛠 See the Toolbox resource, Setting Replication Scope for a New Application Directory Partition by Using NTDSUtil.
5. Verify the creation of both application partitions and that replication has occurred.	▪ Ensure that the application directory partition is created on both domain controllers. 🛠 See the Toolbox resource, Using DNScmd to Verify Creation of Application Partitions.
6. Create stub zones between nwtraders.msft and your forest. The stub zone should be stored in the AppPartition.local application directory partition.	▪ Create the stub zone on only one of the domain controllers in your domain. ▪ The primary name server for the nwtraders.msft domain is london.nwtraders.msft (192.168.*x*.200). 🛠 See the following Toolbox resources: ▪ DNS Security Features in Windows Server 2003 ▪ Creating a DNS Stub Zone ▪ Active Directory Partitions and Replication Scope
7. To ensure that the stub zone is configured correctly, test name resolution for a host in the zone. Ensure that zone transfers are successful for the stub zones.	▪ Ensure that the zone transfers for the stub zone are successful by asking the instructor to create a new Name Server (NS) record in the NWTraders.msft zone. 🛠 See the Toolbox resource, Using Repadmin /syncall /P to force DNS Replication.

Lab E-mail 1

From: Dale Sleppy

To: Systems Engineers

Sent: Mon Sep 08 10:39:00 2003

Subject: Implementing DNS and Active Directory

We have reviewed the DNS design that you developed earlier and have accepted most of it. Here is what we want to you to implement.

You will need to ensure that the following tasks are performed:

- Coordinate between the administrators at headquarters and those in the branch offices to use the Advanced DC Promo process to install the member server in the branch office as another domain controller for your domain.

- After you have installed Active Directory and DNS, move the promoted domain controller from the Domain Controllers organizational unit to the child organizational unit under it called Branch Office.

- Install DNS on the branch office domain controller in your domain, and configure DNS on both domain controllers to meet the following needs:

  - Ensure that each domain controller is using its own DNS server for resolution.

  - Configure the Active Directory DNS zones to ensure that they are available on all domain controllers that are also DNS servers in the domain, and so that DNS zone transfers are optimized.

  - Implement conditional forwarding to contoso.msft on both DNS servers.

  - Ensure that the _MSDCS zone for your forest is highly available.

  - Ensure that, unless otherwise specified, whenever a DNS zone is created all DNS servers in the domain should receive that zone data automatically.

Additionally, Northwind Traders has plans for merging some of our smaller subsidiary companies into our Active Directory in the near future. In anticipation of this merger:

- Create a new DNS forward lookup zone for fourthcoffee.msft on the Main Office DNS server.

- Create a new DNS forward lookup zone for wingtiptoys.msft on the Branch Office DNS server.

As you configure your domain controllers and DNS servers, test their functionality to identify any problems as soon as possible.

Thanks,

Dale Sleppy

MCSE, MCSA, MCT, CNE, CCNA,

Network +, Server +, Security +, CISSP

Managing Network Engineer

Northwind Traders, Inc.

Lab E-mail 2

From: Dale Sleppy

To: Systems Engineers

Sent: Mon Sep 08 10:40:49 2003

Subject: Custom Application and Replication

We have a new development project that will require a new DNS zone for testing purposes. The name of this DNS zone will be AppPartition.local.

We want this new DNS zone to be secured in Active Directory, but we do not want to unnecessarily replicate the DNS zone to all domain controllers. We would like it replicated only between the headquarters and branch office servers to ensure fault tolerance while minimizing overhead due to replication. After you create the application partition, you will create a stub zone for the NWtraders.msft DNS zone and store it in the new application partition.

Additionally, to ensure that we are taking full advantage of all of the Windows Server 2003 Active Directory features, I would like you to raise the forest and domain functionality levels as high as possible for our networking environment.

Dale Sleppy

MCSE, MCSA, MCT, CNE, CCNA,

Network +, Server +, Security +, CISSP

Managing Network Engineer

Northwind Traders, Inc.

Lab Discussion

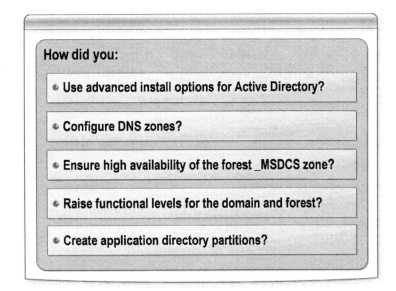

How did you:

- **Use advanced install options for Active Directory?**

- **Configure DNS zones?**

- **Ensure high availability of the forest _MSDCS zone?**

- **Raise functional levels for the domain and forest?**

- **Create application directory partitions?**

Discuss with the class how you implemented the solutions in the preceding lab.

- What account or accounts did you use to perform the tasks in the lab?

- How did you make use of the advanced install options for Active Directory?
    - When would you use this in your organization?
    - When would you not use this in your organization?
    - What are limitations of the advanced options for installation?

- In this lab, you configured three different types of DNS zones: Active Directory domain DNS zones, stub zones, and forward lookup zones. How did you configure the DNS zones on your DNS servers?
    - How did you configure the replication scope for each zone? Why did you configure the replication scope this way?
    - What are the implications for managing DNS zones as Active Directory integrated zones rather than as standard primary and secondary zones?

- How did you ensure high availability of the forest _MSDCS zone?
    - Why is high availability for this zone so important for Active Directory?
    - Under what circumstances would you not use ForestDNSZones to store the _MSDCS zone?
    - What are the advantages of using ForestDNSZones to store the _MSDCS zone?

- How did you raise the domain and forest functional level to that of Windows Server 2003?
    - How did you confirm that the domain and forest had been raised to the intended levels?
    - What are the implications of raising the domain and forest functional levels?
    - Under what circumstances would you raise your domain and forest functional levels to that of Windows Server 2003?
    - Under what circumstances would you not raise your domain and forest levels to that of Windows Server 2003?

- How did you create the application directory partitions?
  - How did you confirm that the application directory partition was created?
  - Under what circumstances would you use an application directory partition?
  - Under what circumstances would an application directory partition not be an appropriate choice?

Best Practices

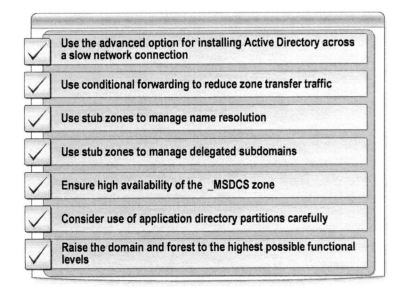

- ✓ Use the advanced option for installing Active Directory across a slow network connection
- ✓ Use conditional forwarding to reduce zone transfer traffic
- ✓ Use stub zones to manage name resolution
- ✓ Use stub zones to manage delegated subdomains
- ✓ Ensure high availability of the _MSDCS zone
- ✓ Consider use of application directory partitions carefully
- ✓ Raise the domain and forest to the highest possible functional levels

---

- *Use the advanced option for installing Active Directory across a slow network connection.* Back up the system state data on a domain controller, copy the backup to portable media, send the media to the remote location, and then restore the backup and use it to install Active Directory. The only changes that will be replicated across the slow network connection to the new domain controller will be the changes made since the backup was created. If you do send the backup media to a remote office, you must ensure that the media is securely sent, stored, and disposed of in a secure manner. The backup media contains the entire Active Directory database for your organization and must be kept secure.

- *Use conditional forwarding to reduce zone transfer traffic.* Without conditional forwarding, you have two choices for resolving DNS names in a forest with multiple namespaces: Configure one server with a secondary copy of all DNS zones and configure all other servers to forward to that server, or configure each of the DNS servers with a secondary zone for each of the DNS zones in the forest. Both options could result in a great deal of zone transfer traffic. With conditional forwarding, you can configure each of the DNS servers to forward DNS queries to the appropriate DNS servers for each zone. This means that there is very little zone transfer traffic, and no single DNS server must respond to all DNS queries.

- *Use stub zones to manage name resolution across multiple namespaces.* You can configure a stub zone on each DNS server for all other namespaces in the company. Each DNS server will then have a current list of all of the name servers for each DNS zone. This means that there will be very little zone transfer traffic, and each DNS server can very efficiently resolve DNS queries.

- *Use stub zones to manage delegated subdomains.* When you set up a delegated subdomain, you must enter the Internet Protocol (IP) address of all of the name servers in the delegated domain. If that list of name servers changes—for example, if one of the name servers is removed from the network—you must manually update the delegation record. You can use a stub zone to automate the process of keeping the name server list updated. To do this, configure the stub zone on the DNS server, and also configure a delegation record for the stub zone.

- *Ensure high availability of the _MSDCS zone*. Configuring the _MSDCS zone for high availability is required for Active Directory to function correctly. This is particularly important if the network environment includes multiple domains and DNS servers in multiple locations. The resource records in the _MSDCS zone will rarely change, so there will be very little replication traffic associated with this zone. The only circumstance in which configuring the _MSDCS zone for high availability by using the ForestDNSZones is not an option is when you are not using Windows Server 2003 to provide DNS services, or if you cannot store the DNS information in Active Directory. In these circumstances, you must develop an alternate strategy to ensure that the _MSDCS zone is available to all DNS clients in the forest.

- *Consider the use of application directory partitions carefully*. Application partitions are most useful for applications that read the directory frequently, or in situations where application data must be available across multiple locations. However, applications that write large amounts of information to the application directory partition will create a significant load on a domain controller. If the application directory partition is replicated to multiple servers, many changes will result in a great deal of replication traffic. Evaluate each application deployment scenario carefully before implementing an application directory partition.

- *Raise the domain and forest to the highest possible functional levels*. To gain the maximum benefits of Windows Server 2003 Active Directory, the domain and forest must be running at the highest functional level. Therefore, raise the domain and forest functional levels as soon as possible. However, raising the functional level is an irreversible process, so you must be absolutely sure that you will never need to install a pre-Windows Server 2003 domain controller in your domain. Functional levels are based only on domain controllers in the domain or forest. You can deploy clients in a domain or forest without any impact on the functional level.

Unit 5: Troubleshooting TCP/IP, Name Resolution, and Group Policy

---

Contents

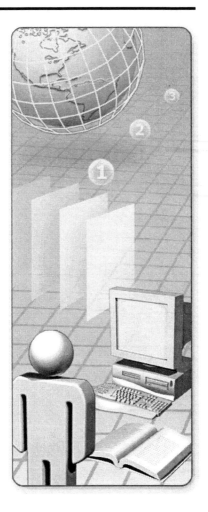

Overview

- **New and Enhanced Troubleshooting Tools**
- **Lab: Troubleshooting TCP/IP, Name Resolution, and Group Policy**
- **Lab Discussion**
- **Best Practices**

In this unit, you will learn about and practice using some of the new and enhanced troubleshooting tools offered with Microsoft® Windows Server™ 2003. The unit will focus particularly on new and improved tools that are part of the new suite of command line and troubleshooting tools for Windows Server 2003.

Objectives

After completing this unit, you will be able to use new or enhanced Windows Server 2003 features to:

- Diagnose and resolve issues related to Domain Name System (DNS) services.
- Troubleshoot Group Policy.
- Diagnose and resolve client computer configuration issues.
- Troubleshoot network connectivity issues.

New and Enhanced Troubleshooting Tools

* Enhanced debug logging
* DNSLint.exe
* DNSCmd.exe
* Netsh.exe
* DCGPOfix.exe

One of the important Windows Server 2003 achievements is providing multiple tools for each server management task. Where possible, graphical user interface (GUI), command line, and Microsoft Windows® Management Instrumentation (WMI) and scripting tools are available for a given task, allowing administrators to select the most convenient or effective tool for the current situation. With this comprehensive set of command line tools, administrators can more efficiently troubleshoot network and server problems by automating key tasks and permitting remote execution.

Some of the key tools include:

- *DNS Debug Logging*. Allows you to log and record packet direction, contents, and other options for general troubleshooting and debugging of a server.

- *DNSLint.exe*. Helps you diagnose common DNS name resolution issues by diagnosing potential causes of "lame delegation" and other related DNS problems, verifying a user-defined set of DNS records on multiple DNS servers, and verifying DNS records specifically used for Microsoft Active Directory® replication.

- *DNScmd.exe*. Allows you to view the properties of DNS servers, zones, and resource records. You can also use DNScmd.exe to modify properties, create and delete zones and resource records, and force replication events between DNS server physical memory and DNS database and data files.

- *Netsh*. Displays or modifies the network configuration of a computer that is currently running.

- *DCGPOfix.exe*. Restores the default Group Policy objects (GPOs) to their original state just after initial installation. It is useful if the default GPOs have been changed to a point at which they have damaged their domains and deleted portions of the Sysvol.

Note For details about using each command line tool and the available options and switches, refer to Windows Server 2003 Help and Support.

 Important

- To use DNSLint to perform an /ad query, you must have Active Directory permissions to perform a Lightweight Directory Access Protocol (LDAP) query. Authenticated users can perform this task.

- To use DNSCmd you must be an Administrator or Server Operator.

If you are an authenticated user, you can use Netsh.exe, but you will need additional permissions, depending on the Netsh context you trying to configure. For example, if you were trying to run **netsh dhcp add server london20.nwtraders.msft 192.168.1.222**, you would need to be an Enterprise Admin.

Lab: Troubleshooting TCP/IP, Name Resolution, and Group Policy

In this lab, you will:

- Configure a DNS server to log client/server communication
- Interpret a DNS Debug Log
- Troubleshoot name resolution and generate a report
- Troubleshoot network connectivity issues and generate a report
- Troubleshoot GPO problems

Objectives

After completing this lab, you will be able to:

- Configure a DNS server to log client/server communication.
- Interpret a Debug Log.
- Troubleshoot name resolution and generate a report.
- Troubleshoot network connectivity issues and generate a report.
- Troubleshoot GPO problems.

Estimated time to complete this lab: **60 minutes**

Toolbox Resources

If necessary, use one or more of the following resources to help you complete this lab:

- New DNS Debug Logging Options
- DNS Debug Log
- DNS Troubleshooting Tools
- DNSLint Interpretation
- Netsh Syntax
- How to Use Netsh
- Using DCGPOfix

Exercise 1
Troubleshooting TCP/IP, Name Resolution, and Group Policy

In this exercise, you will investigate the issues raised in your manager's e-mail message and use the new features of Windows Server 2003 to resolve those issues.

The following table lists specific tasks for you to accomplish. As you complete each task, note your observations and recommendations so that you can include them in your e-mail and discuss them with the class. When you need additional information, use your Toolbox resources.

 Important Ensure that you use an account with the lowest level of administrative permissions required for each task in this exercise.

Tasks	Supporting information
1. Configure a DNS server to log communication between the DNS server and a computer with the IP address 192.168.5.153 to determine why the client is unable to record it's A record in DNS. *Other Options by FILTER packets by IP Address*	■ Navigate to the **Debug Logging** tab in the **Server Properties** dialog box within the DNS console for the server that you want to log, and then set up logging. ■ You will not actually run this log because in the lab setting, there would be no network traffic to log. Instead, you will view a similar log in Task 2. *Incoming (do not use Outgoing)* See the Toolbox resource, New DNS Debug Logging Options.
2. Interpret the debug log and note your observations and recommendations so that you can include them in your e-mail. *Cant send SERV fail 8285*	■ Since there was no network traffic to log in Step 1, you will look at a sample log similar to a small portion of what might have been produced in a real-world setting. ■ Review the 2210A MOD 5 DNS_debug_log.txt file in the Network Files folder to determine the source of the error. See the Toolbox resource, DNS Debug Log. *root hints / ... missing*
3. Troubleshoot name resolution to contoso.msft, generate a report, and note your interpretation of the problem and any recommendations for resolving it.	■ The log generated is simulating a computer named Server2 attempting to update it's A records in DNS. See the following Toolbox resources: *DNSLint contoso.msft 192.168.1.200* ■ DNS Troubleshooting Tools ■ DNSLint Interpretation

DNS root server

Ansend recordadd contoso.msft www a 192.168.5.153

dnslint /d contoso.msft /s 192.168.1.201

(*continued*)

Tasks	Supporting information
4. Troubleshoot Windows Internet Naming Service (WINS) name resolution issues, identify problems, generate a report, and note your recommendations. *Netsh diag ping WINS*	▪ With Microsoft Windows 2000, you used Netdiag. ▪ In Windows Server 2003, use Netsh to explore command line syntax that will allow you to diagnose server issues. ▪ Identify the diagnostic method that enables you to run a report. See the following Toolbox resources: ▪ How to Use Netsh and Run a Report (hypercam) ▪ Netsh Syntax
5. Identify the default domain GPO problem on your domain controller and note your findings and recommendations.	▪ Attempt to edit the default domain policy. ▪ If possible, fix the GPO problem. See the Toolbox resource, Using DCGPOfix.

Lab E-mail

From: Dale Sleppy

To: Systems Engineers

Sent: Mon Sep 08 11:42:45 2003

Subject: Unsolved network issues

The network administrator that was here before you had some network problems that he was unable to solve. I thought you might want to look into these issues to see if you can resolve them. As you know, there are lots of tools and resources at your disposal to aid in troubleshooting.

Here is a listing of the issues that the previous administrator was unable to solve:

1. A client with the IP address 192.168.5.153 has not been able to register its A record into DNS. We need to diagnose and log information on the DNS server both to and from this client only.

2. Name resolution to contoso.msft is no longer working. I would like you to generate a diagnostic report from your computer. Send me your interpretation of the problem and any recommendation that you have to fix it.

3. There have been some issues with your server resolving records from the Microsoft Windows® Internet Name Service (WINS) server. I would like you to use Netsh to resolve this problem and then run a diagnostic report to summarize potential problems with the network setup.

4. We have always had problems with the default domain GPO. No matter what setting that we populate on the default domain policy, nothing seems take effect. Preliminary tests have not determined the problem. Because we do not have any reliable backups to restore the original default domain GPO, I would like you to troubleshoot the problem, research solutions, and restore the original GPO structure on your domain controller, if necessary.

E-mail me with your responses, thanks.

Dale Sleppy

MCSE, MCSA, MCT, CNE, CCNA,

Network +, Server +, Security +, CISSP

Managing Network Engineer

Northwind Traders, Inc.

Lab Discussion

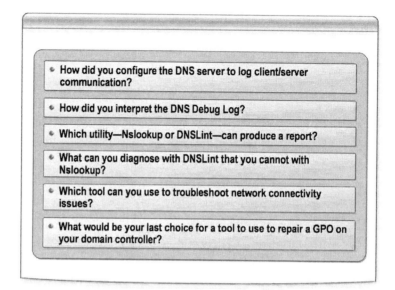

* How did you configure the DNS server to log client/server communication?

* How did you interpret the DNS Debug Log?

* Which utility—Nslookup or DNSLint—can produce a report?

* What can you diagnose with DNSLint that you cannot with Nslookup?

* Which tool can you use to troubleshoot network connectivity issues?

* What would be your last choice for a tool to use to repair a GPO on your domain controller?

- What account or accounts did you use to complete the tasks in the lab?
- How did you configure the DNS server to log client/server communication?
    - Where should you store your logs?
    - What size should the logs be?
    - What debugging options did you set?
    - Why are the default settings for all logging options disabled?
    - How did you interpret the Debug Log?
    - What do the entries in the log mean?
    - What was the source of the error?
- Which utility—Nslookup or DNSLint—can produce a report?
- What can you diagnose with DNSLint that you cannot with Nslookup?
- Which tool can you use for troubleshooting network connectivity issues?
- What would be your last choice for a tool to use to repair a GPO on your domain controller?

Best Practices

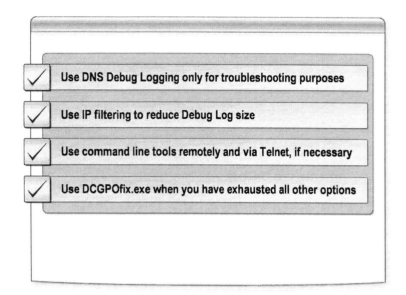

- ✓ **Use DNS Debug Logging only for troubleshooting purposes**
- ✓ **Use IP filtering to reduce Debug Log size**
- ✓ **Use command line tools remotely and via Telnet, if necessary**
- ✓ **Use DCGPOfix.exe when you have exhausted all other options**

- *Use DNS Debug Logging only for troubleshooting purposes.* Debug Logging is resource intensive, so you should only use it when necessary to troubleshoot a problem.

- *Use IP filtering to reduce Debug Log size.* By limiting logging to one IP address or a small set of them, you minimize the resources required to build the log.

- *Use command line tools remotely and via Telnet*, if necessary. Windows Server 2003 supports headless server operation, so with command line administration you can control computers remotely without a keyboard, mouse, or monitor.

- *Use DCGPOfix.exe when you have exhausted all other options.* When you use DCGPOfix.exe, it restores the default GPOs to their original installation states. This of course deletes all custom settings that have been applied since the domain controller installation.

 Note If you use command line tools remotely via Telnet, remember that Telnet is not a secure protocol. Telnet communications can be intercepted—allowing a hacker to obtain access to user names, passwords, and other types of sensitive information. You should take further steps to ensure that Telnet communication is secure, like configuring an IP Security (IPSec) policy for Telnet traffic.

Unit 6: Planning and Implementing Multiple Forests in Active Directory

Contents

trust manual in 2000

Novell cant merge 2 AD's

Overview

- **Planning for a Multi-Forest Environment**
- **Renaming a Domain Controller**
- **Lab: Planning and Implementing Multiple Forests**
- **Lab Discussion**
- **Best Practices**

There are a number of new features in Microsoft® Windows Server™ 2003 that you can use to simplify your task of planning for a multiple forest environment in Microsoft Active Directory®. In this unit, you will use some of the new features of Windows Server 2003 to help you plan and implement multiple forests in Active Directory, which include:

- Forest trust.
- Disabling security identifier (SID) filtering.
- Implementing selective authentication across forest trusts.
- Routing name suffixes across forests.
- Collision detection.
- Renaming domain controllers.
- InetOrgPerson object.

Objectives

After completing this unit, you will be able to:

- Evaluate the need for SID filtering, selective authentication, resolving naming conflicts, and routing name suffixes in a multi-forest environment.
- Establish forest trusts.

Planning for a Multi-Forest Environment

Forest trust
- Allows all domains in one forest to (transitively) trust all domains in another forest, through a single trust link

Disabling SID filtering
- Allows SID history to be used when user migrations are done between forests

Implementing selective authentication
- Allows selective permissions on each computer and resource to which you want users in the second forest to have access

Routing name suffixes between forests
- Manages how authentication requests are routed across Windows Server 2003 forests that use forest trusts

Collision detection
- Ensures that each name suffix is routed to a single forest

Windows Server 2003 provides the following features that will help you plan and implement security and resource access in a multiple-forest environment:

- Forest trust
- Disabling SID filtering
- Selective authentication for a forest trust
- Routing name suffixes across forests
- Collision detection

 Important To create or modify the properties of a forest trust, you need to be a member of the Enterprise Admins group.

Forest Trust

Windows Server 2003 supports the existing Windows 2000 types of trusts and introduces a new type of trust, the *forest trust*. This feature includes:

- *Forest trust*. Allows all domains in one forest to (transitively) trust all domains in another forest, through a single trust link between the two forest root domains.

  - Forest trust is not transitive at the forest level across three or more forests. If forest A trusts forest B, and forest B trusts forest C, this does not create any trust relationship between forests A and C.

  - Forest trusts can be one-way or two-way.

- *Trust management*. Simplifies the creation of all types of trust links, especially forest trusts; a new property page to manage the trusted namespaces associated with forest trusts.

- *Trusted namespaces*. Route authentication and authorization requests for security principals whose accounts are maintained in a trusted forest.

  - The Domain, User Principal Name (UPN), Service Principal Name (SPN), and SID namespaces that a forest publishes are automatically collected when a forest trust is created and refreshed by the Active Directory Domains and Trusts user interface (UI).

  - A forest is trusted to be authoritative for the namespaces that it publishes, on a first-come, first-served basis, as long as the namespaces do not collide with trusted namespaces from existing forest trust relationships.

Disabling SID Filtering on Forest Trusts

This feature enables you to turn off SID filtering on forest trusts, which allows SID history to be used when user migrations occur between forests. You should use this feature only when SID history is needed in a migration scenario. When you use this feature, your forest becomes susceptible to SID spoofing attacks by the administrator of the trusted forest. Therefore, use this feature only when trusted forest administrators can be expected not to indulge in such attacks.

Implementing Selective Authentication for a Forest Trust

You can use Active Directory Domains and Trusts to determine the scope of authentication between two forests that are joined by a forest trust. You can set selective authentication differently for outgoing and incoming forest trusts. With selective authentication, you can make flexible, forest-wide access control decisions. If you decide to set selective authentication on an incoming forest trust, you need to manually assign permissions on each computer and resource to which you want users in the second forest to have access.

Routing Name Suffixes Between Forests

Name suffix routing is a mechanism that you can use to manage how authentication requests are routed between Windows Server 2003 forests that are joined together by forest trusts. To simplify administration of authentication requests, when a forest trust is initially created, all unique name suffixes are routed by default.

Forests can contain multiple unique name suffixes, and all children of unique name suffixes are routed implicitly. If a forest trust exists between two forests, you can use the name suffixes that do not exist in one forest to route authentication requests to a second forest. When a new child name suffix (*.child.Nwtraders.com) is added to a unique name suffix (*.Nwtraders.com), the child name suffix will inherit the routing configuration of the unique name suffix to which it belongs. Any new unique name suffixes that are created after a forest trust has been established will be visible in the forest trust **Properties** dialog box after you verify the trust. However, routing for those new unique name suffixes will be disabled by default.

When a duplicate name suffix is detected, the routing for the newest name suffix will be disabled by default. You can use the forest trust **Properties** dialog box to manually prevent authentication requests for specific name suffixes from being routed to a forest.

Collision Detection

When two Windows Server 2003 forests are linked by a forest trust, there is a possibility that unique name suffixes existing in one forest may collide with unique name suffixes located in the second forest. Collision detection ensures that each name suffix will only be routed to a single forest.

When a name suffix in a forest conflicts with a new forest trust partner, or when a name suffix in an existing forest trust conflicts with a new forest trust partner, the name will be disabled in the new trust. For example, a conflict will occur if one forest is named widgets.com and the second forest is named sales.widgets.com. Despite the name suffix conflict, routing will still work for any other unique name suffixes in the second forest. If Active Directory Domains and Trusts detects a name suffix conflict with a trust partner domain, access to that domain from outside the forest may be denied. However, access to the conflicted domain from within the forest will function normally.

Renaming a Domain Controller

* **Renaming a domain controller**
  * Allows restructuring of your network for organizational and business needs
  * Makes management and administrative control easier
* **To rename a domain controller**
  * If the domain functional level is set to Windows Server 2003, use:
    - Control Panel
    - Netdom.exe

Windows Server 2003 provides the ability to rename domain controllers without demoting them, which allows you to make changes to your forest structure and namespace as the needs of your organization change. You can rename a domain controller to:

- Restructure your network for organizational and business needs.

- Make management and administrative control easier.

- Resolve network basic input/output system (NetBIOS) naming conflicts.

You can rename domain controllers by using the Control Panel or Netdom.exe.

 Note Renaming a domain controller may cause it to become temporarily unavailable to users and computers. Also, certain services, such as the certification authority (CA), rely on a fixed computer name. Check first to determine whether any services of this type are running on the domain controller.

 Important To rename a domain controller, you need to be a member of the Domain Admins group.

Renaming Domain Controllers

If the domain functional level is set to Windows Server 2003, you can rename a domain controller by using the Control Panel or Netdom.exe.

To rename a domain controller by using the Control Panel, perform the following steps:

1. Run the **System** utility for Control Panel.

2. In the **System Properties** dialog box, on the **Computer Name** tab, click **Change**.

3. When prompted, confirm that you want to rename the domain controller.

4. Type the full computer name, including the primary Domain Name System (DNS) suffix, and then click **OK**.

To rename a domain controller by using Netdom.exe, perform the following steps:

1. Open a command prompt.

2. Type **netdom computername** *CurrentComputerName* **/add:***NewComputerName*

 Note This command will update the service principal name (SPN) attributes in Active Directory for this computer account and register DNS resource records for the new computer name. The SPN value of the computer account must be replicated to all domain controllers for the domain, and the DNS resource records for the new computer name must be distributed to all of the authoritative DNS servers for the domain name. If the updates and registrations have not occurred prior to removing the old computer name, then some clients may be unable to locate this computer by using the new or old name.

3. Verify that the new computer name has been properly registered in DNS and in Active Directory. Type **netdom computername** *CurrentComputerName* **/verify**

4. Assign the alternate name that was added in step 2 as the primary name for the domain controller. Type **netdom computername** *CurrentComputerName* **/makeprimary:***NewComputerName*

5. Restart the computer.

6. From a command prompt, type **netdom computername** *NewComputerName* **>/remove:***OldComputerName*

Additional Information For information about renaming domains, see Domain Rename Steps.ppt under **Additional Reading** on the Web page on the Student Materials compact disc.

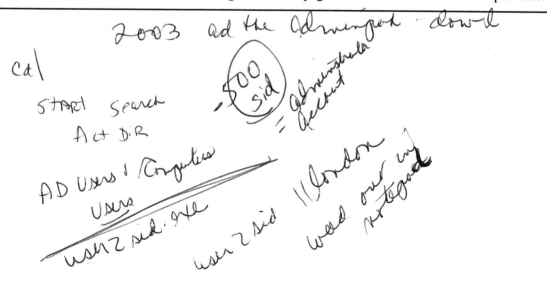

Lab: Planning and Implementing Multiple Forests

In this lab, you will:

- Verify functionality levels for a forest trust
- Implement a two-way forest trust
- Implement selective authentication
- Evaluate effects of a rogue or coerced administrator in another forest
- Verify and test the forest trust
- Research and test domain controller renaming
- Research interoperability between Active Directory and other directory services

After completing this lab, you will be able to:

- Verify domain and forest functionality levels for a forest trust.
- Implement a two-way forest trust.
- Implement selective authentication between forests.
- Evaluate effects of a rogue or coerced administrator in another forest.
- Verify and test a forest trust.
- Rename a domain controller.
- Evaluate interoperability between Active Directory and other directory services.

Estimated time to complete this lab: **60 minutes**

Toolbox Resources

If necessary, use one or more of the following Toolbox resources to help you complete this lab:

- Configuring Selective Authentication in a Two-Way Forest Trust
- Creating a Two-Way Forest Trust
- Verifying a Trust Relationship Using Netdom.exe
- Verifying Functional Level by Using ADSIEdit
- Verifying Functional Level by Using LDP
- What Is SID Spoofing?
- What Is the InetOrgPerson Object?

Exercise 1
Implementing a Two-Way Forest Trust

In this exercise, you will:

- Verify domain and forest functionality levels for a forest trust.

- Implement a two-way forest trust.

- Implement selective authentication between forests.

- Research and implement methods to protect your forest from a rogue or coerced administrator in a trusted forest.

- Verify and test the trust.

- Research domain controller renaming, and test it on your domain controller by renaming your domain controller with your first name. *Other LDAP dir serv AD usersclient access inter org pers allows you to do this. The others*

 Important Ensure that you use an account with the lowest level of administrative permissions required for each task in this exercise.

Tasks	Supporting information

1. Verify that the prerequisites for creating a forest trust are met.	▪ To verify that the prerequisites have been met, consider the following methods:
	  • DNS console
	  • Active Directory Domains and Trusts
	  • The LDP tool
	  • ADSIEdit
	
	▪ Decide which approach is the best solution, and then implement it. Be prepared to explain your answer.
	
	See the following Toolbox resources:
	  ▪ Prerequisites for Creating a Forest Trust
	  ▪ Verifying Functional Level by Using LDP
	  ▪ Verifying Functional Level by Using ADSIEdit

2. Create a two-way forest trust between your forest and the contoso.msft forest. When you create the trust, select the **Forest-wide authentication** option.	▪ This task can only be performed on one domain controller in your forest. Coordinate with the administrator of the other domain controller in your forest to accomplish this task.
	
	▪ The user name and password for the account in the contoso.msft forest that has the appropriate privileges to create a forest trust are as follows:
	  • User name: **Administrator**
	  • Password: **P@ssw0rd**
	
	See the Toolbox resource, Creating a Two-Way Forest Trust.

(*continued*)

Tasks	Supporting information
3. Ensure that users who are members of the Domain Users global group from contoso.msft are the only users who can be authenticated by the domain controllers in headquarters and the branch office.	■ To prevent users from being authenticated by other computers in your forest, implement selective authentication on your forest trust. By implementing selective authentication, you can grant users permission to be authenticated by the specific computers in your forest that contain resources that you want them to access. 　See the Toolbox resource, Configuring Selective Authentication in a Two-Way Forest Trust.
4. Identify possible security concerns with rogue or coerced administrators from contoso.msft, and take the necessary steps (if any) to protect your forest.	See the Toolbox resource, What Is SID Spoofing?
5. Verify that the trust relationship has been correctly established, and then test the ability to log on with an account from the contoso.msft forest.	■ You can use Netdom.exe or Active Directory Domains and Trusts to verify the trust relationship. ■ Log on to your domain controller by using the following user name and password from the contoso.msft domain: 　● User name: **DiazB** 　**Important** The Default Domain Controller policy for all domains in the classroom was configured during setup to grant the **Allow Log on Locally** user right to the Authenticated Users special group. If this setting has been inadvertently revoked, you will not be able to log on as **DiazB**. 　See the Toolbox resource, Verifying a Trust Relationship Using Netdom.exe.
6. Rename your domain controller by adding the number 1 to the end of the current name.	■ To implement this requirement, you could use the following methods: 　● Netdom.exe 　● The System utility in Control Panel ■ Decide which approach is the best solution, and then implement it. Be prepared to explain your answer.

Exercise 2
Evaluating Interoperability Between Different Directory Services

In this exercise, you will research issues on interoperability between Active Directory and other Lightweight Directory Access Protocol (LDAP)-compatible directory services.

Tasks	Supporting information
1. Research the InetOrgPerson object feature of Windows Server 2003 that facilitates interoperability between Active Directory and other LDAP-compatible directory services.	■ To implement this requirement, consider the following issues: • LDAP user account type • Ability of various account types to be security principles in Active Directory ■ Complete your research, and then reply to the e-mail. See the Toolbox resource, What Is the InetOrgPerson Object?

Lab E-mail 1

From: Dale Sleppy

To: Systems Engineers

Sent: Mon Sep 08 11:53:51 2003

Subject: Acquisition of Contoso Ltd.

As mentioned earlier, we have had plans to acquire contoso.msft. Those plans have been realized, and we now need to plan our strategy for secure authentication between the two organizations. In the Network Trust diagram, I have outlined the resources that need to be accessed. We need to ensure that users who are members of the Domain Users global group from contoso.msft are the only users who can be authenticated by the domain controllers in the headquarters and the branch office. I am giving you the task of planning and implementing this. I want you to perform the following tasks:

- Implement a two-way forest trust between your forest and the contoso.msft forest.

- Ensure that users who are members of the Domain Users global group from contoso.msft are the only users who can be authenticated by the domain controllers in the headquarters and the branch office.

- Identify possible security concerns with rogue or coerced administrators from contoso.msft, and take any necessary steps to protect our forest.

- After the trust has been established verify that the trust relationship has been correctly established, and test the ability to log on with an account from the contoso.msft forest. You can use the contoso\DiazB user account with our default password to perform this test.

- There are naming conflicts between our domain controllers and the domain controllers in the contoso.msft forest. I understand that there is a way to rename domain controllers in Windows Server 2003. I want you to rename your domain controller by adding the number 1 to the end of the current name.

Thanks,

Dale Sleppy

MCSE, MCSA, MCT, CNE, CCNA,

Network +, Server +, Security +, CISSP

Managing Network Engineer

Northwind Traders, Inc.

Lab E-mail 2

From: Dale Sleppy

To: Systems Engineers

Sent: Mon Sep 08 11:54:52 2003

Subject: Interoperability with Netscape Directory services

We also need to assess our ability to interoperate with Netscape Directory Services in the future. I would like you to research the InetOrgPerson object feature of Windows Server 2003 that facilitates interoperability between Active Directory and other LDAP-compatible directory services. After you have completed your research, please send me an e-mail explaining the various options.

Thanks,

Dale Sleppy

MCSE, MCSA, MCT, CNE, CCNA,

Network +, Server +, Security +, CISSP

Managing Network Engineer

Northwind Traders, Inc.

Lab Discussion

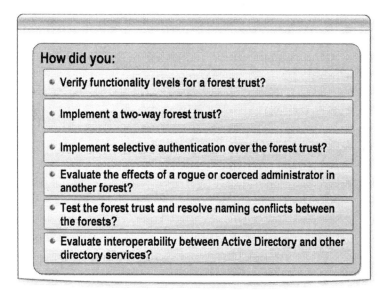

How did you:

- Verify functionality levels for a forest trust?

- Implement a two-way forest trust?

- Implement selective authentication over the forest trust?

- Evaluate the effects of a rogue or coerced administrator in another forest?

- Test the forest trust and resolve naming conflicts between the forests?

- Evaluate interoperability between Active Directory and other directory services?

Discuss with the class how you implemented the solutions in the preceding lab.

- What account or accounts did you use to complete the tasks in the lab?
- How did you verify functionality levels for a forest trust? Did you use: *ADSI edit*
  - The DNS console?
  - Active Directory Domains and Trusts?
  - The LDP tool?
  - ADSIEdit?
  - A different solution?
- How did you implement a forest trust? Did you:
  - Use Netdom.exe?
  - Use Active Directory Domains and Trusts? ✓
  - Successfully validate the trust? *yes .*
  - Use a different solution?
- How did you ensure that users from a forest can only be authenticated by specific computers? Did you:
  - Use Netdom.exe or Active Directory Domains and Trusts to configure selective authentication for the trust?
  - Assign the **Allowed to Authenticate** permission to individual users, to a group, or to everyone?
  - Use a different solution?

- How did you evaluate the effects of a rogue or coerced administrator in another forest? Did you use:
  - Netdom.exe to enable SID filtering?
  - Netdom.exe to disable SID filtering?
  - A different solution?
- How did you test the forest trust and resolve naming conflicts between the forests? Did you use:
  - Active Directory Domains and Trusts to validate the trust?
  - Netdom.exe to verify the trust?
  - Control Panel to rename your domain controller?
  - Netdom.exe to rename your domain controller?
  - A different solution?
- How did you evaluate interoperability between Active Directory and other directory services? Did you plan to use:
  - InetOrgPerson accounts? — open Standard
  - Microsoft MetaDirectory Services? miis (single sign on product)
  - A third-party solution?
  - A different solution?

Best Practices

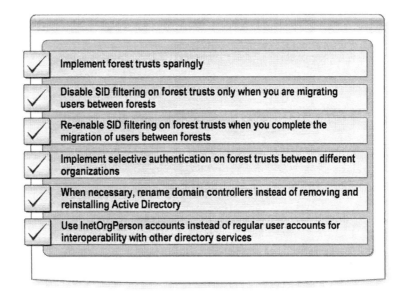

- *Implement forest trusts sparingly.* Implement forest trusts only when users from multiple domains in one forest need to access resources in multiple domains in another forest. If users from a single domain in one forest need access to resources in a single domain in another forest, consider implementing an external trust instead of a forest trust.

- *Disable SID filtering on forest trusts only when you are migrating users between forests.* Always leave SID filtering enabled on a forest or external trust unless you are migrating users between domains over the trust relationship. If you are migrating users between forests that have a trust relationship, and the migrated users need to access resources in the forest from which they were migrated, disable SID filtering on the forest trust.

- *Re-enable SID filtering on forest trusts when you complete the migration of users between forests.* If you disable SID filtering on a forest trust because you are migrating users between forests, consider re-enabling SID filtering when all users and resources have been migrated and discretionary access control lists (DACLs) on resources have been updated.

- *Implement selective authentication on forest trusts between different organizations.* This will enable you to specify the users in the other forest who can access certain servers in your forest.

- *When necessary, rename domain controllers instead of removing and reinstalling Active Directory.* When restructuring domains, or when duplicate domain controller names exist in a multiple forest environment, consider renaming one of the domain controllers instead of removing and reinstalling Active Directory.

- *Use InetOrgPerson accounts instead of regular user accounts for interoperability with other directory services.* When you plan to use Active Directory and one or more additional LDAP-compatible directory services on your network, consider implementing user accounts as InetOrgPerson accounts to facilitate interoperability between the directories.

Unit 7: Using Group Policy in Windows Server 2003 to Deploy and Restrict Software

Contents

Overview

- **Reasons for Managing a User Environment**
- **Limiting User Software and Script Activity**
- **Creating Software Distribution Packages**
- **Lab: Implementing Group Policy to Deploy and Restrict Software**
- **Lab Discussion**
- **Best Practices**

Group Policy in Microsoft® Windows Server™ 2003 includes tools and resources that enable you to better control the user environment on your network. By using Group Policy, you can limit the use of unproductive and potentially harmful software. You can also create software distribution packages that will completely install an assigned application when a user logs on. You can also use Windows Management Instrumentation (WMI) filters to restrict the computers where the application is installed.

Objectives

After completing this unit, you will be able to:

- List reasons for controlling a computer user's environment.
- Create a software restriction policy.
- Deploy software so that an application is completely installed at user logon.
- Use WMI filters to restrict the application of Group Policy objects (GPOs).

Reasons for Managing a User Environment

Why manage a user environment?
- Consistency for users
- Ease of management for administrators
- Protection against potentially harmful software

Many organizations recognize the benefits of managing the computing environment for end users within the company. These benefits include:

- *Consistent user environment.* Central management and control of the user computing environment ensures that all users benefit from an optimized desktop and a consistent experience.

- *Ease of management for administrators.* Much of an organization's total cost of ownership (TCO) of desktop computers is spent on managing computers after purchase. This includes the initial cost of configuring and deploying the workstations and the ongoing cost of troubleshooting and reconfiguring computers and applications after the deployment. A centralized means for managing large numbers of user desktops can significantly decrease your TCO.

- *Protection against potentially harmful software.* Viruses and other dangerous applications can cause millions of dollars worth of damage to an organization's network. By managing the installation of software and enacting policies to prevent dangerous applications from running, you can greatly reduce these costs.

Roaming profile thing

Limiting User Software and Script Activity

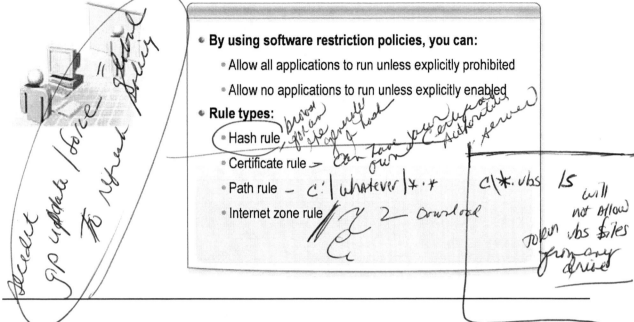

* **By using software restriction policies, you can:**
  * Allow all applications to run unless explicitly prohibited
  * Allow no applications to run unless explicitly enabled
* **Rule types:**
  * Hash rule
  * Certificate rule
  * Path rule
  * Internet zone rule

Software restriction policies are new options for managing user environments in Windows Server 2003. You use software restriction policies to control what applications are allowed to run on user workstations and to prevent dangerous or unwanted applications from running on those workstations.

When you configure software restriction policies, you must configure a restriction level for the policy. By default, a software restriction policy allows all applications to run unless they are explicitly prohibited. You can modify this setting so that the default configuration allows no applications to run unless they are explicitly enabled.

Rule Types

After you have configured the default setting for the software restriction policies, you can configure exceptions to the policy by using four different types of rules that apply to hashes, certificates, paths, and Internet zones.

You configure software restriction policies by using Group Policy, and you apply them just as you would any other Group Policy. For example, you could create a domain level GPO that configures a strict software restriction policy for all users in the domain except members of administrative groups. Or you could create a less restrictive policy for all users in the domain, and then apply a very restrictive policy only for computers or users that require more strict controls. By creating an organizational unit structure that enables these options, and then applying the GPO to the appropriate containers in Microsoft Active Directory®, you can implement a flexible and powerful software restriction policy.

 Important To create and modify software restriction policies, you need to be a member of the Domain Admins group.

Creating Software Distribution Packages

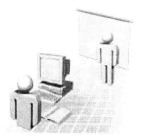

* **Install at logon**
  * You can configure Group Policy to install software when a user logs on
* **WMI filters**
  * You can use filters to regulate application of GPOs based on criteria you specify

---

Windows Server 2003 Active Directory includes enhanced options for installing software using Group Policy. The two primary enhancements are the option to completely install an assigned application when a user logs on, and the option to use WMI filtering to restrict the application of Group Policy.

Install at Logon

In Microsoft Windows® 2000 Active Directory, an assigned application is only installed when a user activates the installation by using the **Start** menu, by extension activation, or through **Add or Remove Programs** in Control Panel. The option to perform a complete installation in Windows Server 2003 means that users may have to wait for the application to install when they log on, but they will not have to wait for the installation when they want to use the application. At logon, users can only completely install applications that are assigned to them; published applications are not installed until the user activates the installation.

WMI Filtering

WMI filtering limits the application of a GPO to computers that meet criteria that you specify. For example, you may want to install an application only on computers that have at least 256 megabytes (MB) of random access memory (RAM) and at least 500 MB of free space on the hard disk. To do this, you can create a GPO that will install the application and then link a WMI filter with the necessary criteria to the GPO. When the GPO is applied, the filter is evaluated on the target computer. If the WMI filter evaluates to false, the GPO is not applied. If the WMI filter evaluates to true, the GPO is applied. In this case, if the computer does not meet the minimum requirements defined in the WMI filter, the application would not be installed.

 Important To create and modify software restriction policies, you need to be a member of the Domain Admins group.

Lab: Implementing Group Policy to Deploy and Restrict Software

In this lab, you will:

- Create a software distribution package
- Use WMI filters to control the application of Group Policy
- Generate a Group Policy settings report

After completing this lab, you will be able to:

- Create a software distribution package.
- Use WMI filters to control the application of Group Policy.
- Generate a Group Policy settings report.

Estimated time to complete this lab: **60 minutes**

S/w restriction policies

Toolbox Resources

- Configuring Software Restriction Policies
- Detailed Steps: Configuring Software Restriction Policies
- How Software Restriction Policies Are Applied
- Resultant Set of Policy Reports
- WMI Filters
- Using WMI Filtering
- Install at Logon Feature
- Using Group Policy Results and Group Policy Modeling

Admin script editor

ASE

Exercise 1
Configuring Group Policy for Software Restrictions and for Installing Software

In this exercise, you will begin by creating an organizational unit for the Headquarters and Branch Office administrator accounts. All of the modifications that you make in this exercise will be performed on the administrator's organizational unit for your office. You will then configure Group Policy settings to implement a software restriction policy. You will also configure Group Policy settings that will completely install an assigned software package the next time that the user logs on. Finally, you will create a report of the Group Policy settings that are applied to your user account and your computer account.

 Important Ensure that you use an account with the lowest level of administrative permissions required for each task in this exercise.

Tasks	Supporting information
1. Create a new organizational unit in your domain for your office. This organizational unit will be used for all of the user accounts for the office administrators.	▪ The Headquarters administrator should create an organizational unit named HeadQuarters Admins organizational unit. Move your regular user account and your administrator user accounts into the organizational unit. ▪ The Branch Office administrator should create an organizational unit named BranchOffice Admins organizational unit. Move your regular user account and your administrator user accounts into the organizational unit.
2. If you do not have the Group Policy Management Console (GPMC) installed, install it now.	▪ To install the GPMC, run \\London\Setup\gpmc.msi.

3. Create a software restriction policy that prevents users from running Microsoft Outlook® Express. Implement the policy on the Admins organizational units created in the previous step.	▪ Use the **GPUpdate** command to force an immediate refresh of the Group Policy after you create the policy.
	▪ You could implement this policy by using:
	  • Hash rule.
	  • Path rule.
	  • Registry path rule.
	▪ What option did you chose? Be prepared to explain your choice.
	
	See the following Toolbox resources:
	  ▪ Configuring Software Restriction Policies
	  ▪ Detailed Steps: Configuring Software Restriction Policies
	  ▪ How Software Restriction Policies Are Applied

(continued)

Tasks	Supporting information
4. Configure the software restriction policy so that only Microsoft Visual Basic® (.vbs) scripts that have been signed with the corporate certificate will run on user desktops.	■ The code signing certificate is located in the Network Files folder. Copy the certificate to the root of drive C and use that location when configuring the certificate rule. ■ How can you configure the software policies so that only the .vbs scripts that are signed by the corporate certificate will run? See the following Toolbox resources: 　■ Configuring Software Restriction Policies 　■ Detailed Steps: Configuring Software Restriction Policies 　■ How Software Restriction Policies Are Applied
5. Ensure that the software restriction policy that prevents users from running Outlook Express is successfully applied.	■ Can you run Outlook Express with your non-administrative user account? Can you run Outlook Express with your administrative account?
6. Configure a software installation package that: 　• Assigns all of the administrative tools available in Adminpak.msi to administrators. The tools should be installed when the administrators log on with either their regular accounts or their administrator accounts. 　• Completely installs the administrative tools on the administrator's computer the next time that the administrator logs on. 　• Only installs the administrative tools if the computer has more than 2 GB of free space.	■ Configure the software installation package for each of the Admins organizational units. ■ The Adminpak.msi file is located on the Windows Server 2003 installation CD in the I386 folder. ■ When creating a software distribution package, you must use a Universal Naming Convention (UNC) path to a shared folder. Create a shared folder on the server that you are administering, and then copy the Adminpak.msi file into that folder. ■ 2 gigabytes (GB) is equal to 2147483648 bytes. See the following Toolbox resources: 　■ Install at Logon Feature 　■ WMI Filters 　■ Using WMI Filtering

(continued)

Tasks	Supporting information
7. Ensure that the administrative tools are completely installed on the administrators' computers.	■ Are the administrative tools completely installed for your administrative account? For your non-administrative account? ■ A subset of administrative tools is automatically installed on all domain controllers. This subset does not include administrative tools like Dynamic Host Configuration Protocol (DHCP) or Windows Internet Naming Service (WINS) unless these services are installed on the server. You can check to determine whether the administrative tools installed correctly by checking for the availability of these additional administrative tools.

8. Generate a report that provides detailed information about the Group Policies being applied to your non-administrative user account and the domain controller to which you are logging on. Send the report as an e-mail message to your supervisor.	■ To create a Group Policy report, you can use:
	  • Group Policy Modeling Wizard.
	  • Group Policy Results Wizard.
	  • gpresult.exe.
	■ What tool did you use? Be prepared to explain your choice.
	
	See the following Toolbox resources:
	  ■ Resultant Set of Policy Reports
	  ■ Using Group Policy Results and Group Policy Modeling

Lab E-mail

From: Dale Sleppy

To: Systems Engineers

Sent: Mon Sep 08 13:14:35 2003

Subject: Software Deployment and Restrictions

Based on our discussion, please use Group Policy to support the following recommendations:

- Limit the use of unproductive and potentially harmful software. We want all users to use Outlook as their e-mail client at all times, so you need to remove the ability for users to run Outlook Express. Additionally, you must prevent any Visual Basic scripts from running that have not been signed by our company developers with a certificate from our corporate CA.

- Create a software distribution package that will install Adminpak.msi for administrators the next time they log on to their computers. This program should not be installed on computers that have less than 2 gigabytes (GB) of free disk space.

After you have implemented the Group Policy, submit a report showing what the effective group policy is for client computers and users in your organizational unit.

Thanks,

Dale Sleppy

MCSE, MCSA, MCT, CNE, CCNA,

Network +, Server +, Security +, CISSP

Managing Network Engineer

Northwind Traders, Inc.

Edit Computer + User Config'g

Group Pol Loopback merge or replace

Lab Discussion

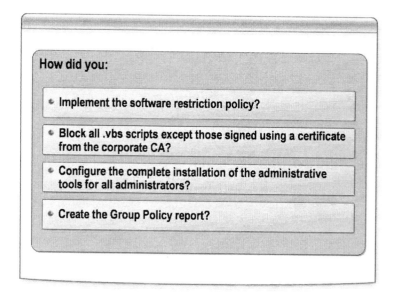

How did you:

- Implement the software restriction policy?
- Block all .vbs scripts except those signed using a certificate from the corporate CA?
- Configure the complete installation of the administrative tools for all administrators?
- Create the Group Policy report?

Discuss with the class how you implemented the solutions in the preceding lab.

- What account or accounts did you use to complete the tasks in the lab?
- How did you implement the software restriction policy to prevent users from running Outlook Express?
  - How did you choose your option?
  - Why did you not choose one of the other available options?
  - What software restriction policies would be most useful in your organization?
  - How would you implement the policies?
- How did you block all .vbs scripts from running except for the scripts signed using a certificate from the corporate certification authority (CA)?
  - Is there any other way that you could have achieved this?
  - How did you choose your option?
- How did you configure the complete installation of the administrative tools for all administrators?
  - What are some benefits of performing a complete install rather than allowing the user to initiate the installation of the application?
  - What are some disadvantages?
- How did you create the Group Policy report on all of the Group Policy settings that are applied to the user and computer accounts?
  - How did you choose the tool to use?
  - Under what circumstances would you use the other available tools?

Best Practices

> ✓ Create a separate GPO for software restriction policies
>
> ✓ Restart in Safe Mode if you experience problems with software restriction policies
>
> ✓ For highly-secured workstations, configure the default setting for all applications to Disallowed
>
> ✓ Test new software restriction policies thoroughly
>
> ✓ Consider bandwidth implications of complete installs of applications
>
> ✓ Use WMI filtering only when necessary
>
> ✓ Use WMI filtering with Group Policy Modeling and Results tool

- *Create a separate GPO for software restriction policies.* Processing multiple policies can slow down the logon process for users, but maintaining the separate policy makes it much easier to isolate and troubleshoot problems with the software restriction policy. Using a separate policy also means that you can disable software restriction policies in an emergency without disabling other settings in your policy.

- *Restart the computer in Safe Mode if you experience problems with software restriction policies.* Software restriction policies do not apply when Windows is started in Safe Mode.

- *For highly-secured workstations, configure the default setting for all applications to* **Disallowed**. When you define a default setting of **Disallowed**, no software will run on the computer except for software that has been explicitly allowed. This can be useful for computers that must be secured as much as possible, or when few applications are ever going to run on the computer.

- *Test new software restriction policy settings thoroughly.* Use test environments before applying new policy settings to your production environment. When you do deploy the policies to the production computers, consider creating a test organizational unit and moving a pilot group of users into it first before rolling out the deployment to the entire network.

- *Consider bandwidth implications of complete installs of applications.* If you deploy a large application to many users, and those users all log on at approximately the same time, it may take a long time for the application to install. In general, choose the complete installation option only if you are deploying the application to a smaller number of users who will log in at the same time, or if the application is fairly small.

- *Use WMI filtering only when necessary.* WMI filters provide flexible options in managing the application of Group Policies to client workstations. However, implementing several WMI filters can result in a very complex environment and can degrade performance when a user logs on. Instead, manage the application of Group Policy through your organizational unit design and by security group filtering. Use WMI filters only when there is no other option.

- *Use WMI filtering with the Group Policy Modeling and Results tool.* You can create WMI filters that restrict the application of GPOs by using almost any hardware or software related value on the target computer. However, WMI filtering does not by default provide any reporting tool that reports which GPOs were blocked on which computers. If you use WMI filtering extensively, you should use the Group Policy Modeling and Results tool frequently to determine the effective Group Policy settings on client computers.

Workshop Evaluation

Your evaluation of this workshop will help Microsoft understand the quality of your learning experience.

At a convenient time before the end of the workshop, please complete a workshop evaluation, which is available at http://www.CourseSurvey.com.

Microsoft will keep your evaluation strictly confidential and will use your responses to improve your future learning experience.

Unit 8: Using Group Policy in Windows Server 2003 to Set Advanced Security Settings

Contents

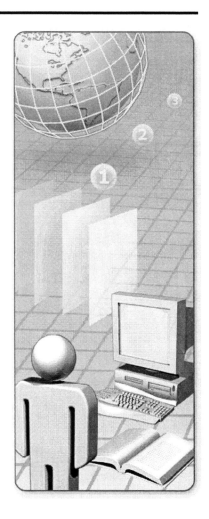

Overview

* **Wireless Policies**
* **Configuring EFS Settings**
* **Lab: Advanced Security Settings in Group Policy**
* **Lab Discussion**
* **Best Practices**

Microsoft® Windows Server™ 2003 provides many improved features for creating a strong and flexible security system; these features include file encryption and wireless network policies. In this unit, you will learn how to set these advanced security options by using Group Policy.

Objectives

After completing this unit, you will be able to:

■   Configure wireless network settings.

■   Configure a user environment.

■   Apply Encrypting File System (EFS) enhancements.

Wireless Policies

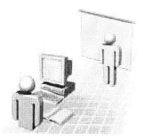

* **Security mechanisms for wireless networks**
    * 802.11, 802.1X
* **To create wireless network policies:**
    * Use Active Directory Users and Computers or the Group Policy Management Console
* **Use wireless policies to specify:**
    * Whether clients can use Windows to configure wireless network connection settings
    * Whether to enable 802.1X authentication for wireless network connections
    * Preferred wireless networks to which clients can connect

---

Wireless networking technologies provide convenience and mobility, but they also introduce security risks on your network. For example, unless authentication and authorization mechanisms are implemented, anyone who has a compatible wireless network adapter can access a network. Without encryption, wireless data is sent in plaintext, so anyone within sufficient distance of a wireless access point (AP) can detect and receive all data sent to and from a wireless AP. Many access points ship with nonsecured default configurations that broadcast Service Set Identifiers (SSIDs) without encryption enabled. Access points ship with a default SSID that is known to attackers.

 Important To create a wireless policy, you need to be a member of the Domain Admins group.

The following security mechanisms enhance security over wireless networks:

- *SSID*. In 802.11, each client uses a SSID as a limited security mechanism. Only clients and access points that have the same SSID can communicate with each other. There is no additional authentication, so anyone who can determine the SSID can potentially connect to the wireless network.

- *Wired Equivalent Privacy (WEP)*. In WEP, each client uses a pre-shared key to encrypt data that is sent over the wireless network. The keys used for WEP encryption can be either 40 or 104 bits in length. WEP encryption is vulnerable to attack because the keys are not changed dynamically during each session; instead, they must be manually changed by an administrator on each access point and client computer.

- *802.1X.* 802.1X was designed to create a more secure connection mechanism.

  - It can work on both wired and wireless local area networks (WLANs). 802.1X authenticates the connection before allowing the client to access the network. 802.1X uses Extensible Authentication Protocol-Transport Layer Security (EAP-TLS) protocol to authenticate the connection.

  - 802.1X also requires Remote Authentication Dial-In User Service (RADIUS). RADIUS is a way to forward authentication requests from a RADIUS client to a RADIUS server that can check the authorization and tell the client to allow or deny access. In 802.1X, the wireless access point functions as a RADIUS client.

 Additional Information For more information about 802.11 and 802.1X security mechanisms, see Wlandeploy.doc under **Additional Reading** on the Web page on the Student Materials compact disc.

To enhance the deployment and administration of wireless networks, Windows Server 2003 provides Active Directory Users and Computers or the Group Policy Management console to configure a Group Policy object (GPO) that enforces wireless network restrictions.

When you use Group Policy to define wireless network policies, you can specify:

- Whether clients can use Microsoft Windows® to configure wireless network connection settings.

- Whether to use 802.1X authentication for wireless network connections.

- The preferred wireless networks to which clients can connect.

 Note For enhanced security, 802.1X authentication is enabled by default in the Windows Server 2003 family of products. The Institute of Electrical and Electronics Engineers (IEEE) 802.1X authentication provides authenticated access to 802.11 wireless networks and to wired Ethernet networks. IEEE 802.1X authentication provides support for EAP security types so that users can employ authentication methods such as certificates.

Configuring EFS Settings

* **Enhancements to EFS in Windows Server 2003**

  * Stronger encryption algorithms with larger keys
    - Can also use 3DES 168-bit encryption
    - Use FIPS compliant algorithms for encryption, hashing, and signing

  * Authorization of multiple users to share encrypted files
    - You must be a member of the administrators group or be the user that encrypted the file
    - Users must have a file encryption certificate assigned to them, or should have previously encrypted a file

  * Encryption of offline files
    - Enable the **Encrypt the Offline Files** cache property

---

EFS allows users to store their on-disk data in encrypted format. Encryption is the process of converting data into a format that other users cannot read. Once a user has encrypted a file, the file automatically remains encrypted whenever the file is stored on disk.

In Windows Server 2003, EFS includes the following enhancements:

- *Stronger encryption algorithms with larger keys.* Windows Server 2003 uses Advanced Encryption Standard (AES), which uses a 256-bit key for encryption. The encryption in Windows Server 2003 is more complex and stronger than in Windows 2000. However, you can use a Group Policy security setting that allows computers running Windows XP and Windows Server 2003 to use the 3DES 168-bit encryption. To set the 168-bit encryption:

 Important To configure EFS encryption options by using Group Policy, you need to be a member of the Domain Admins group.

- Navigate to **Windows Settings\Security Settings\Local Policies\Security Options**.

- Select the **Use FIPS compliant algorithms for encryption, hashing, and signing** option.

- *Authorization of multiple users to share encrypted files.* To authorize sharing of encrypted files by multiple users, you must be a member of the administrators group or be the user that encrypted the file. You can retain the security of file encryption while allowing specific users access to your encrypted files. To allow access to your encrypted files:

  - Right-click the encrypted file, and then click **Properties**.

  - Click the **General** tab (if it is not already selected), and then click **Advanced**.

  - Click **Details**, and then click **Add**.

  - Select the users to whom you want to provide access to the encrypted file.

 Important Before you can allow users to share an encrypted file, the users must have a file encryption certificate assigned to them, or they should have previously encrypted a file.

- *Encryption of offline files through EFS, which enables protection of locally cached documents.* Offline files reside on a user's hard disk, and they are stored in a local cache on the computer. Encrypting this cache enhances security on a local computer. To enable encryption of offline files:

  - In the Group Policy Object Editor, expand **Computer Configuration**, **Administrative Templates**, and **Network**, and then click **Offline Files**.

  - In the details pane, double-click **Encrypt the Offline Files cache**.

  - Select the **Enabled** option to enable offline file encryption.

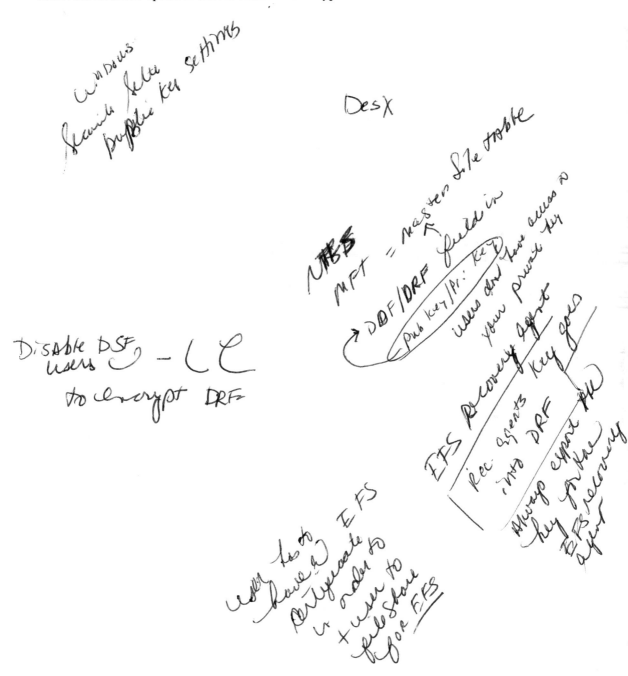

Lab: Advanced Security Settings in Group Policy

In this lab, you will:

- Create a wireless network access policy
- Implement firewall settings
- Implement EFS restrictions
- Generate a Group Policy settings report

After completing this lab, you will be able to:

- Create a wireless network access policy.
- Implement firewall settings.
- Implement EFS restrictions.
- Generate a Group Policy settings report.

Estimated time to complete this lab: **60 minutes**

Toolbox Resources

If necessary, use one or more of the following Toolbox resources to help you complete this lab:

- Creating a Wireless Policy
- Configuring Advanced EFS Options
- Disabling EFS in Windows 2000, Windows XP, and Windows 2003
- Prohibiting the Use of Internet Connection Firewall

Exercise 1
Configuring Security Settings by Using Group Policy

In this exercise, you will:

- Create three new organizational units for your location.

- Create a new GPO that implements a wireless network policy for your location.

- Configure a GPO to implement firewall settings.

- Configure a GPO to prevent the use of EFS encryption on all computers running Windows 2000 in your domain.

- Configure a GPO to prevent the use of EFS encryption on all domain controllers in your location.

- Configure a GPO to enforce advanced EFS encryption options.

- Use Windows Server 2003 reporting tools to prepare a report on your computer that shows the effective results of your policies on computers in your location.

 Important Ensure that you use an account with the lowest level of administrative permissions required for each task in this exercise.

Tasks	Supporting information

1. In your student domain, create the following organizational units:
 	
▪ Headquarters
   • HQ Computers
   • HQ Users
▪ Branch Office
   • BO Computers
   • BO Users
	▪ The administrator of the Headquarters domain controller should create the organizational units for Headquarters, and the administrator of the Branch Office domain controller should create the organizational units for the Branch Office.
	
	▪ To create these organizational units, consider the following methods:
	   • Dsadd.exe
	   • Active Directory Users and Computers ✓
	   • Active Directory Service Interfaces (ADSI) scripting
	
	▪ Decide which approach is the best solution, and then implement it. Be prepared to explain your answer.

(continued)

Tasks	Supporting information

2. Create a new GPO for Computers organizational unit for your location (where location is either Headquarters or Branch Office) that implements a wireless network access policy that fulfills the following:
 	
• Connects to WLAN_*Location* SSID
 	
• Uses IAS01 and IAS02 as the RADIUS servers
 	
• Uses a trusted root certification authority (CA) of the VeriSign Trust Network
 	
• Supplies WEP Support with automatic supplying of keys
 	
• Offers 802.1X authentication using EAP-MSCHAP v2
 	
• Enables fast reconnection between multiple access points
	▪ To create a new wireless network access policy, consider the following methods:
	
	   • Active Directory Users and Computers
	
	   • Group Policy Management console
	
	▪ Decide which approach is the best solution, and then implement it. Be prepared to explain your answer.
	
	See the Toolbox resource, Creating a Wireless Policy.

3. Ensure that mobile users in your location have firewalls in place when they are away from the office. Also ensure that users do not have firewalls in place while they are connected to your network.	▪ To implement this requirement, you could:
	
	   • Establish a company policy requiring users to configure a firewall on their mobile computers when they are not connected to the company's network.
	
	   • Have technicians configure the firewall settings on the mobile computers, and use Group Policy to prevent the use of firewalls on your network.
	
	▪ Decide which approach is the best solution, and then implement it. Be prepared to explain your answer.
	
	See the Toolbox resource, Prohibiting the Use of Internet Connection Firewall.

(continued)

Tasks	Supporting information
4. Prevent the use of EFS encryption on all computers running Windows 2000 in your domain.	▪ You can perform this task on only one domain controller in your forest. Coordinate with the administrator of the other domain controller in your forest to accomplish this task. ▪ Since this task involves modifying the Default Domain Policy, you should use the Group Policy Management Console (GPMC) to back up the Default Domain Policy before you modify it. To back up the Default Domain Policy, in GPMC, right-click the **Default Domain Policy** GPO, and then click **Backup**. See the Toolbox resource, Disabling EFS in Windows 2000, Windows XP, and Windows 2003.
5. Use Group Policy to prevent the use of EFS encryption on all domain controllers in your location.	▪ To implement this requirement, you could: • Place all domain controllers from your location in a separate organizational unit and configure Group Policy for that organizational unit to disable EFS. • Create a new GPO for the organizational unit that corresponds to your location, and configure it to disable EFS. Next, you could create a security group that has all of the domain controllers in your location as members, and then configure security filtering for the GPO so that the GPO only applies to members of the security group. • Create a new GPO for the organizational unit that corresponds to your location, and configure it to disable EFS. Then create a Windows Management Instrumentation (WMI) filter that queries for domain controllers, and link it to the GPO. Use the following query: `Select DomainRole FROM Win32_ComputerSystem WHERE (DomainRole = "4" OR DomainRole = "5")` **Note** Use WMI filters sparingly on GPOs that apply to domain controllers. Excessive use of WMI filters on GPOs that will be processed by domain controllers can cause performance degradation on the domain controllers, because Group Policy is updated every five minutes. ▪ Decide which approach is the best solution, and then implement it. Be prepared to explain your answer. See the Toolbox resource, Disabling EFS in Windows 2000, Windows XP, and Windows 2003.

(*continued*)

Tasks	Supporting information

6. Use Group Policy to configure advanced EFS settings for computers in your location to achieve the following:
 	
• When EFS is used on Windows Server 2003 member servers and Windows XP clients, use 168-bit encryption.
 	
• Mobile users who use offline files will use EFS for the client side cache.
	All of the required settings are located under **Computer Configuration**.
	
	The option to configure 168-bit encryption is located in:
	
	■ **Windows Settings\Security Settings\Local Policies\Security Options**
	   • System cryptography: Use Federal Information Processing Standard (FIPS) compliant algorithms for encryption, hashing, and signing.
	
	■ **Administrative Templates\Network\Offline Files**
	   • Encrypt the Offline Files Cache.
	
	See the Toolbox resource, Configuring Advanced EFS Options.

7. Use Windows Server 2003 reporting tools to prepare a report on your computer that shows the effective results of your policies on computers in your location.	■ To implement this requirement, you could:
	   • Use the Group Policy Results feature of the GPMC.
	   • Use the Resultant Set of Policy snap-in.
	   • Use the Gpresult.exe command line utility.
	   • Use the resultant set of policy feature in the Windows Server 2003 Help and Support Center. To access this, open Help and Support, click **Tools**, click **Help and Support Center Tools**, click **Advanced System Information**, and then click **View Group Policy settings applied**.
	
	■ Decide which approach is the best solution, and then implement it. Be prepared to explain your answer.

Lab E-mail

From: Dale Sleppy

To: Headquarters administrators; Branch Office administrators

Sent: Mon Sep 08 13:39:29 2003

Subject: Security Settings

To ensure optimal security and availability for the wireless network in your OU, implement Group Policies that meet the following criteria:

- A wireless network policy in your OU for the WLAN_organizational unit wireless network. This wireless network policy should allow for the following:
  - Connect to WLAN_organizational unit SSID
  - Provide WEP Support with automatic supplying of keys
  - Use 802.1X authentication using EAP-MSCHAP v2
  - Enable fast reconnection between multiple access points
- Ensure that mobile users have firewalls in place when they are away from the office. Also ensure that users do not have firewalls in place while they are connected to your network.
- Configure EFS settings in your OU to achieve the following:
  - EFS should not be allowed on domain controllers or computers running Windows 2000.
  - When EFS is used on Windows Server 2003 member servers and Windows XP clients, use 168-bit encryption.
  - Mobile users who use offline files will use EFS for the client side cache.

After you have implemented Group Policy, submit a report showing what the effective Group Policy is for your OU and client computers.

Thanks,

Dale Sleppy

MCSE, MCSA, MCT, CNE, CCNA,

Network +, Server +, Security +, CISSP

Managing Network Engineer

Northwind Traders, Inc.

Lab Discussion

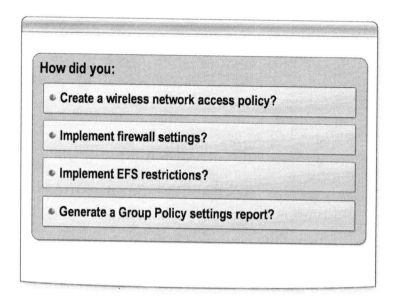

How did you:

- Create a wireless network access policy?
- Implement firewall settings?
- Implement EFS restrictions?
- Generate a Group Policy settings report?

Discuss with the class how you implemented the solutions in the preceding lab.

- What account or accounts did you use to complete the tasks in the lab?
- How did you create a new wireless policy? Did you use:
  - Active Directory Users and Computers?
  - The GPMC?
  - A different solution?
- How did you implement firewall restrictions and requirements? Did you:
  - Establish a company policy requiring the use of Internet Connection Firewall (ICF)?
  - Use Group Policy to prevent the use of ICF on your corporate network?
  - Manually configure ICF on all mobile computers?
  - Use a different solution?
- How did you prevent the use of EFS encryption on all computers running Windows 2000 in your domain? Did you:
  - Use Group Policy to prevent the use of EFS?
  - Remove the default recovery agent from the Default Domain Policy?
  - Use a different solution?
- How did you prevent the use of EFS on domain controllers in your location? Did you:
  - Use Group Policy to prevent the use of EFS?
  - Place all domain controllers from your location in their own organizational units?
  - Create a new security group with all of your domain controllers as members?
  - Create a WMI filter for domain controllers?
  - Use a different solution?

- How did you configure advanced EFS settings for your location? Did you:
  - Use Group Policy to implement the EFS settings?
  - Manually configure all of the computers in your location?
  - Use a different solution?
- How did you prepare a report showing the effective results of the Group Policy settings that you implemented in this lab? Did you use:
  - The GPMC?
  - The Resultant Set of Policy snap-in?
  - Gpresult.exe?
  - The Help and Support Center?
  - A different solution?

Best Practices

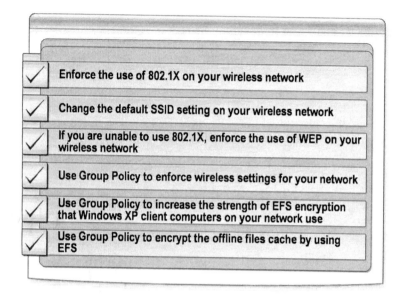

- *Enforce the use of 802.1X on your wireless network.* Implement this requirement if all of the following prerequisites are met:
  - All of the computers on your network that require wireless network access support 802.1X.
  - Your wireless APs support 802.1X.
  - Your network infrastructure includes a RADIUS server.
  - You have the required infrastructure to implement user certificates or smart cards.

- *Change the default SSID setting that your wireless network uses.* Each manufacturer of wireless products uses a default SSID for its products. Network crackers know these SSIDs and use them to find unsecured wireless networks. You should change the SSID from the default to increase the security of your network.

- *(If you are unable to use 802.1X), require the use of Wired Equivalent Policy (WEP) on your wireless network.* If for any reason you are unable to require the use of 802.1X on your wireless network, at least implement WEP encryption for all traffic on your wireless network.

- *Use Group Policy to enforce wireless settings for your network.* Group Policy is the easiest way to configure wireless settings for all of the computers on your network. However, these settings only apply to computers that are running Windows XP or later operating systems.

- *Use Group Policy to increase the strength of EFS encryption that computers running Windows XP on your network use.* If you choose to allow the use of EFS on your network, consider using Group Policy to configure all computers running Windows XP and later to use 3DES (168-bit) encryption for EFS. 3DES is considerably more secure than the Windows XP default, DESX (120-bit) encryption.

- *Use Group Policy to encrypt the offline files cache by using EFS.* If users of mobile computers store confidential company information in their offline files caches, consider using Group Policy to require encryption of offline files. This Group Policy setting only applies to computers running Windows XP or later.

Unit 9: Planning and Implementing Secure Routing and Remote Access

Contents

Overview

- Configuring Routing and Remote Access
- Configuring, Securing, and Monitoring IPSec
- Lab: Planning and Implementing Secure Routing and Remote Access
- Lab Discussion
- Best Practices

This unit provides hands-on experience by using new and enhanced Microsoft® Windows Server™ 2003 Routing and Remote Access services and Internet Protocol security (IPSec) features to implement secure routing and remote access.

Objectives

After completing this unit, you will be able to use Windows Server 2003 features to:

- Plan, implement, and maintain Routing and Remote Access.
- Create and implement an IPSec policy.
- Configure IPSec by using Netsh.
- Set up IPSec policy monitoring.

Configuring Routing and Remote Access

- PPPoE for demand dial connections and broadband Internet connections
- NetBIOS over TCP/IP name resolution proxy
- L2TP/IPSec NAT traversal
- New default behaviors for registering internal and external interfaces
- Routing and Remote Access Server Setup Wizard
- Basic Firewall

Routing and Remote Access in the Windows Server 2003 family provides multiprotocol, local area network (LAN)-to-LAN, LAN-to-wide area network (WAN), virtual private network (VPN), and network address translation (NAT) routing services. It is largely unchanged from Microsoft Windows® 2000 Routing and Remote Access with the exception of some new enhancements:

- *Point-to-Point Protocol over Ethernet (PPPoE).* PPOE is now supported for demand dial connections and broadband Internet connections.

- *Network basic input/output system (NetBIOS) over Transmission Control Protocol/Internet Protocol (TCP/IP) name resolution proxy.* A remote access client can resolve full computer names and NetBIOS names of computers and other resources on a remote network that does not have either a Domain Name System (DNS) or a WINS server configured.

- *Layer Two Tunneling Protocol and IPSec (L2TP/IPSec) NAT traversal.* VPN servers running Windows Server 2003 support L2TP/IPSec traffic that originates from VPN clients behind NATs. For this feature, the client computer must support negotiation of NAT-T traversal in the Internet Key Exchange (IKE) and support User Datagram Protocol (UDP) encapsulation of IPSec packets.

- *New default behaviors for registering internal and external interfaces.* To prevent possible problems when resolving a VPN server's name and accessing services on the VPN server, the Windows Server 2003 Routing and Remote Access service, by default, disables the Dynamic DNS registration for the internal interface and Dynamic DNS and NetBIOS over TCP/IP (NetBT) of the external interface.

- *Routing and Remote Access Server Setup Wizard.* This wizard makes setting up and configuring Routing and Remote Access quick and easy. (For special situations, manual configuration is also available.)

- *Basic Firewall.* Basic Firewall allows you to configure which services and ports are exposed to the public interface if your network has no other firewall software installed.

 Important To create or modify the properties Routing and Remote Access you need to be a member of the Administrators group on the local or remote machine. If the computer is a domain member, the Domain Admins group can perform these tasks.

Configuring, Securing, and Monitoring IPSec

- **IP Security Monitor**
  - Setting up policies
  - Main mode (IKE)
  - Quick mode (PSec)
- **Persistent IPSec policies**

The new IP Security Monitor snap-in provides detailed IPSec policy configuration and active security state. This replaces the Windows 2000 Ipsecmon.exe tool. An IPSec policy consists of sets of main-mode policies, quick-mode policies, main-mode filters, and quick-mode filters (for both transport and tunnel mode). *Active security state* refers to the active main-mode and quick-mode security associations and statistical information about IPSec protected traffic.

- *Setting up policies*. This procedure is largely unchanged from Windows 2000. However, in Windows 2000 and Windows XP, by default all broadcast, multicast, Internet Key Exchange (IKE), Kerberos version 5 protocol, and Resource Reservation Protocol (RSVP) traffic is exempt from IPSec filtering. To significantly improve security, in the Windows Server 2003 family only IKE traffic (which is required for establishing IPSec-secured communication) is exempt from IPSec filtering. All other traffic types are now matched against IPSec filters, and you can configure, block, or permit filter actions specifically for multicast and broadcast traffic.

- *Main mode*. Establishes the Internet Security Association and Key Management Protocol (ISAKMP) Oakley security association.

- *Quick mode*. Establishes the IPSec security association.

- *Persistent IPSec policies*. An administrator can use new functionality with the Netsh IPSec context to ensure availability of IPSec policies.

 Important To create or modify an IPSec policy you need to be a member of the Administrators group on the local or remote machine. If the computer is a domain member, the Domain Admins group can perform these tasks.

 Note You can also use Network Monitor and Netsh to monitor IPSec on Windows Server 2003.

Lab: Planning and Implementing Secure Routing and Remote Access

In this lab, you will:

- Set up a server as both a VPN and a NAT server
- Ensure that only necessary services are exposed to the Internet
- Verify that the public IP address is not registered in DNS
- Configure IPSec with contoso.msft
- Ensure that IPSec policy will be applied if either local or domain IPSec policy is not available
- Set up IPSec monitoring

Objectives

After completing this lab, you will be able to:

- Set up a server running Windows Server 2003 as both a VPN and a NAT server.
- Expose only necessary services to the Internet.
- Ensure that the public IP address of the server running Routing and Remote Access is not registered in DNS.
- Configure secure communications with contoso.msft by using IPSec.
- Configure the IPSec policy to be applied if either the local or domain IPSec policy is not available.
- Set up IPSec monitoring.

Estimated time to complete this lab: **60 minutes**

Toolbox Resources

If necessary, use one or more of the following resources to help you complete this lab:

- Enabling Routing and Remote Access as a VPN and NAT Server
- Services and Ports Tab
- Routing and Remote Access Network Interface Enhancements
- Configuring IPSec to contoso.msft
- Common IPSec Configuration Errors
- Using Netsh to Make Persistent IPSec Policies
- Using the IP Security Monitor Console
- IPSec Monitoring Tools

Exercise 1
Planning and Implementing Secure Routing and Remote Access

In this exercise, you will investigate the issues raised in your manager's e-mail message and use the new features of Windows Server 2003 to resolve those issues.

The following table lists specific tasks for you to accomplish. When you need additional information, use your Toolbox resources.

Note Be sure to enable the second network adaptor for this lab. To reduce the chance of confusion, rename the 192.168.*x.y* adaptor **External** and the second network adaptor **Internal**.

Important Ensure that you use an account with the lowest level of administrative permissions required for each task in this exercise.

Tasks	Supporting information
1. Set up the server as both a VPN server and a NAT server.	▪ Use the 192.168.*x.y* adaptor as the External Interface. ▪ Enable Routing and Remote Access as a VPN and a NAT server. See the Toolbox resource, Enabling Routing and Remote Access as a VPN and NAT Server.
2. Minimize the server's attack profile by confirming that only necessary services are exposed to the Internet.	▪ Confirm that only services that support NAT and VPN are selected. See the Toolbox resource, Services and Ports Tab.
3. Ensure that the public IP address is not registered in DNS.	▪ Verify that no A record exists for the external interface. See the Toolbox resource, Routing and Remote Access Network Interface Enhancements.
4. Configure secure communication with contoso.msft by using IPSec.	▪ Consider any filter rules necessary to ensure secure communication only with 192.168.*x*.201. See the following Toolbox resources: ▪ Configuring IPSec to contoso.msft ▪ Common IPSec Configuration Errors
5. Ensure that the IPSec policy will be applied if either the local or domain IPSec policy is not available.	▪ Consider any filter rules necessary to ensure secure communication only with 192.168.*x*.201. See the Toolbox resource, Using Netsh to Make Persistent IPSec Policies.

(continued)

Tasks	Supporting information
6. Test the IPSec connection by setting up and monitoring IPSec.	■ Consider the following monitoring alternatives, and then choose the best one for this situation: • Microsoft Management Console (MMC) snap-in • Netsh IPSec • Network Monitor See the following Toolbox resources: ■ Using the IP Security Monitor Console ■ IPSec Monitoring Tools

aarong Lanclerode
@ springhouse.
Com

2 nics =
————————————————
1 NAT connected to outside

Lab E-mail

From: Dale Sleppy

To: Systems Engineers

Sent: Mon Sep 08 13:53:28 2003

Subject: Remote Access and VPN Support for Mobile Users

I have just been notified by the Chief Security Officer that, due to recent security concerns, we need to better secure our communication path with contoso.msft. We have decided to implement the following measures to better control and secure resources in our organization:

- Dedicated IPSec communication between YourComputer and contoso.msft.
- A VPN server for mobile users from both contoso.msft and YourComputer.
- An NAT server to allow multiple users to use one public IP address.

We have installed another server running Windows Server 2003 with a separate WAN line connected to it. We need you to configure our IPSec policy, VPN, and NAT by accomplishing the following tasks:

- Set up the server as both a VPN server and an NAT server.
- Confirm that only necessary services are exposed to the Internet to ensure that the attack profile of the server is minimized.
- Verify that the server running Routing and Remote Access does not register the public IP address in DNS.
- Set up an IPSec policy for your organizational unit by using Group Policy to secure a communication path with contoso.msft only.
- Ensure that the IPSec policy will remain in effect, even if the Local or Domain Group Policy is unavailable.
- Test the connection to contoso.msft by setting up and monitoring IPSec.

Thanks,

Dale Sleppy

MCSE, MCSA, MCT, CNE, CCNA,

Network +, Server +, Security +, CISSP

Managing Network Engineer

Northwind Traders, Inc.

Lab Discussion

- Is it easier to set up the server as both a VPN and NAT server by using the wizard or by doing it manually?

- What options are available for Basic Firewall?

- How are IP addresses for servers that run Routing and Remote Access registered?

- Which filters did you use to configure secure communication?

- Should you use persistent policies? What are the implications of using persistent policies?

- How did you set up monitoring to verify communication?

Discuss with the class how you implemented the solutions in the preceding lab.

- What account or accounts did you use to complete the tasks in the lab?

- Is it easier to set up the server as both a VPN server and a NAT server by using the wizard, or by doing it manually? Why?

- What options are available for Basic Firewall? What is the behavior of a stateful firewall?

- How are IP addresses for servers that run Routing and Remote Access registered?

- Which filters did you use to configure secure communication?

- Should you use persistent policies? What are the implications of using persistent policies?

- How did you set up monitoring to verify communication? Where would you look for troubleshooting errors?

Best Practices

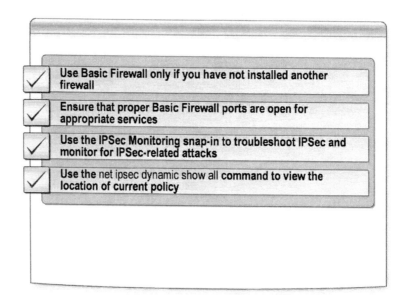

- *Use Basic Firewall only if you have not installed another firewall.* You do not need to use Basic Firewall if your computer has other firewall software installed, but you can use Basic Firewall as another layer of defense in conjunction with another firewall on the network. Basic Firewall can provide protection only for public interfaces. You cannot enable Basic Firewall on a private interface. For Basic Firewall to work correctly with all services and ports on an interface, you must configure each instance of Basic Firewall individually.

- *Ensure that proper Basic Firewall ports are open for appropriate services.* This will ensure that applications running on the server that require unsolicited requests from the Internet will be able to facilitate those requests.

- *Use the IPSec Monitoring snap-in to troubleshoot IPSec and monitor for IPSec-related attacks.* You can use the IPSec Monitoring snap-in as a troubleshooting tool as well as a tool for monitoring for suspected attacks. For example, a large number of quick-mode packets with bad security parameters indexes (SPIs) that are received within a short amount of time might indicate a packet spoofing attack.

- *Use the **net ipsec dynamic show all** command to view the location of the current policy.* This command allows you to quickly see which IPSec policy is applied to the computer.

Workshop Evaluation

Your evaluation of this workshop will help Microsoft understand the quality of your learning experience.

To complete a workshop evaluation, go to http://www.CourseSurvey.com.

Microsoft will keep your evaluation strictly confidential and will use your responses to improve your future learning experience.

Appendix A: Network Files

The following sections contain information from the network files in the Resource Toolkit.

Company Security Policies

Acceptable Use of Administrator and Non-Administrator Accounts

Users who perform administrative functions must undergo a thorough background check before being granted administrative access.

After an administrator has been given authority for a server or domain, the administrator must create the required administrator accounts. The administrator must create all required administrator accounts so that the administrator can always use an account with the lowest level of administrative rights to accomplish any task.

An administrator may only log on with an administrator account to perform a task that will not work by using the Secondary Logon service. For example, it is not possible to run Microsoft® Windows® Explorer by using Run As, so administrators can perform those tasks while logged on as Administrator if their user accounts do not have sufficient rights or permissions. After completing these tasks, the user must immediately log off and log on with a non-administrative account. Failure to comply with these policies can result in disciplinary action.

User Account Naming

Account names are created by using the last name followed by one or more letters of the first name to establish a unique name. Administrator accounts must be created with the same name followed by an underscore character (_) and the type of administrator account, for example, Username_DomAdmin or Username_DNSAdmin.

A complex password must be assigned to all user accounts by default.

Corporate Locations

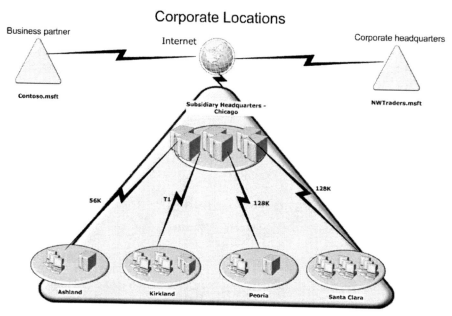

New subsidiary - Your forest

DNS Configuration

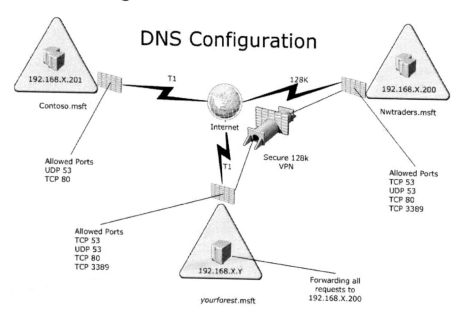

Network Trust

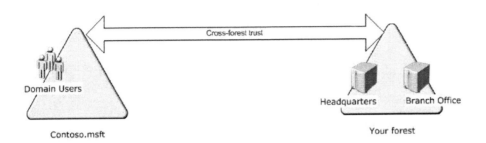

Target State DNS

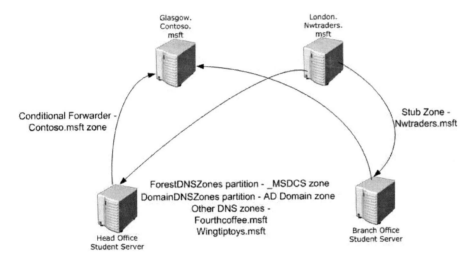

DNS Server Log

DNS Server log file creation at 4/16/2003 5:26:10 PM UTC

Message logging key:

Field #	Information	Values
1	Remote IP	
2	Xid (hex)	
3	Query/Response	R = Response
		blank = Query
4	Opcode	Q = Standard Query
		N = Notify
		U = Update
		? = Unknown
5	[Flags (hex)	
6	Flags (char codes)	A = Authoritative Answer
		T = Truncated Response
		D = Recursion Desired
		R = Recursion Available
7	ResponseCode]	
8	Question Name	

10:26:12 EBC EVENT   The DNS server has started.

10:26:19 F10 PACKET UDP Rcv 192.168.5.153   0058 Q [0001 D NOERROR]
(7)server2(6)adatum(3)com(0)

10:26:20 F10 PACKET UDP Rcv 192.168.5.153   0058 Q [0001 D NOERROR]
(7)server2(6)adatum(3)com(0)

10:26:21 F10 PACKET UDP Rcv 192.168.5.153   0058 Q [0001 D NOERROR]
(7)server2(6)adatum(3)com(0)

10:26:23 F10 PACKET UDP Rcv 192.168.5.153   0058 Q [0001 D NOERROR]
(7)server2(6)adatum(3)com(0)

10:26:27 F10 PACKET UDP Rcv 192.168.5.153   0058 Q [0001 D NOERROR]
(7)server2(6)adatum(3)com(0)

10:26:37 EB8 PACKET UDP Snd 192.168.5.153   0058 R Q [8285 A DR SERVFAIL]
(7)server2(6)adatum(3)com(0)

10:26:38 F10 PACKET UDP Rcv 192.168.5.153   0059 Q [0001 D NOERROR]
(7)server2(6)adatum(3)com(0)

10:26:39 F10 PACKET UDP Rcv 192.168.5.153   0059 Q [0001 D NOERROR]
(7)server2(6)adatum(3)com(0)

10:26:40 F10 PACKET UDP Rcv 192.168.5.153   0059   Q [0001   D   NOERROR]
(7)server2(6)adatum(3)com(0)

10:26:42 F10 PACKET UDP Rcv 192.168.5.153   0059   Q [0001   D   NOERROR]
(7)server2(6)adatum(3)com(0)

10:26:46 F10 PACKET UDP Rcv 192.168.5.153   0059   Q [0001   D   NOERROR]
(7)server2(6)adatum(3)com(0)

10:26:56 EB8 PACKET UDP Snd 192.168.5.153   0059 R Q [8285 A DR SERVFAIL]
(7)server2(6)adatum(3)com(0)

MSM2210BCPWKBK/C90-02044

0 99751 06766 0